The Delft Systems Approach

Hans P.M. Veeke • Jaap A. Ottjes • Gabriël Lodewijks

The Delft Systems Approach

Analysis and Design of Industrial Systems

Springer

Hans P.M. Veeke, PhD
Jaap A. Ottjes, PhD
Gabriël Lodewijks, PhD

Department of Marine and Transport Technology
Faculty of Mechanical, Maritime and Materials Engineering (3ME)
Delft University of Technology
Mekelweg 2
2628 CD Delft
The Netherlands

ISBN 978-1-84800-176-3 e-ISBN 978-1-84800-177-0

DOI 10.1007/978-1-84800-177-0

British Library Cataloguing in Publication Data
Veeke, Hermanus Petrus Maria
The Delft systems approach : analysis and design of industrial systems
1. Industrial management - Mathematical models
I. Title II. Ottjes, Jaap A. III. Lodewijks, Gabriel
658.4'033
ISBN-13: 9781848001763

Library of Congress Control Number: 2008928769

Cover design: eStudio Calamar S.L., Girona, Spain

Printed on acid-free paper

9 8 7 6 5 4 3 2 1

springer.com

To Professor Jan in 't Veld
(1925–2005)

Preface

This book is a tribute to Prof. Jan in 't Veld, who is the real founder of the Delft Systems Approach. Until a few days before he died in 2005, he was working on the English translation of his successful book *Analyse van Organisatie-problemen*. The eighth edition of the book appeared in the Netherlands in 2002; it represents 50 years of experience, discussion and application of a systems approach that was gradually developed and sharply defined by him. About 85,000 copies of the book had been sold by then.

Unfortunately he was not able to finish the translation. The writers of this book are convinced of the value of his approach and decided to finish his work. The chapters of the Dutch book that had already been translated (by A.M.P. O'Brien) are included here with only minor changes made to them (Chapts. 2, 3, 4, 7, 8 and 9). However, we extended the theory with a conceptual view of behaviour modelling, a view that was approved by in 't Veld himself.

Besides this, in 't Veld started his work from a purely managerial viewpoint. The approach was primarily a contribution to management science. We start from the perspective of the engineering world and therefore restrict the approach to the level of operational management. We have often found that the approach contributes to a quick but thorough understanding of operational problems, and it has helped us (and many students) to not just solve the *problems correctly* (an engineering skill) but to identify the *correct problems* (a supposed management skill).

With this first English edition, we hope to expose an international audience to this unique approach that is widely applied in The Netherlands and has received great appreciation, not just from students and academic colleagues, but mostly from managers in practice. They all recognize the challenging statement of in 't Veld: "managers do know what they want to know, but they rarely know what they should know". This statement remains relevant even today, with the rise of powerful information systems and emerging technologies like wireless communication and RFID (Radio Frequency IDentification).

Delft, December 2007

Hans Veeke
Jaap Ottjes
Gabriël Lodewijks

Contents

Chapter 1
Introduction

Abstract. This book is primarily intended for engineers that reach a managerial position in organisations in the industrial sector (production or transport), public and private services, societies, banks, etc., as well as for members of staff who support these functions. In addition, this book is also intended for engineering students who aspire to a managerial or staff position in the future.

Whenever we use the word "manager", we specifically address the engineer who performs a managing function.

1.1 The Purpose of this Book

The purpose of this book is as follows:

- To improve the conception of a design in order to obtain a better match between the expected operation and the real operation of an (intended) industrial system.
- To integrate the structural and behavioural conceptions of a system to be designed.
- To support communication between different specialists, both of whom are involved in the same processes and projects.
- To guide managers in their application of this knowledge to the problems with which they are confronted in practice.
- To communicate knowledge and understanding of the part of systems theory that will enable managers to further improve their performance and/or to reduce their workload.

H.P.M. Veeke, J.A. Ottjes, G. Lodewijks, *The Delft Systems Approach*,

1.2 Theory and Practice

The pace of the development of knowledge and know-how in the organisation sciences, logistics and information technology is rapid. The gap between those who practice these sciences and the practicing manager is however becoming larger rather than smaller. In general, a business management technique has an incubation time of approximately five years before it becomes an integral part of a company's toolkit, whereas the incubation period of basic theory and a new method of thinking is a complete generation.

From the outset of our professional careers, we have campaigned for thinking in product streams and in processes in order to reduce flow time and increase flexibility. Although systems theory emerged as a science in the 1940s, only in the 1980s did it become apparent that systems thinking, thinking in terms of processes and system models, was actually being applied in companies. However, since then, its application has enjoyed rapid expansion. Banks, insurance companies and other service-providing organisations are now implementing this approach, which is resulting in the resolution of many of their problems.

There are many possible reasons for the gap between theory and practice. We believe that there are four main reasons:

1. The difference between finding a solution in theory and in practice
2. The time pressure of management
3. The fashion consciousness of business management science
4. The perception of "importance".

Ad 1. Too many scientists presume that the manager attempts to find the absolute best solution to a problem. In practice, the interest level in doing this is not optimal. The manager is primarily interested in avoiding poor solutions. Should one useable alternative be available that is not too risky then he is satisfied enough to "choose" this solution. He is usually pleased that one such alternative has been found. On the other hand, he is conscious of the personal risks involved, from a self-protection viewpoint. He therefore seeks methods to both solve problems and to analyse the risk elements of the various alternatives presented to him. The chosen alternative is not too important to him so long as it does not have a disastrous result. No one can ever prove that with a different choice, better results would have been achieved. There is actually never a control situation where the circumstances are 100% identical. In addition, a group can realise almost any reasonable organisational decision when the members of the group trust in the decision. A manager is thus only marginally interested in various sophisticated mathematical methods for partial problems, unless the advantage to be had is clearly demonstrable in advance.

Ad 2. Also, the manager never has enough time. Taking a day out to attend a course on new developments is asking a lot. Reading a book, should he do so, is usually of little benefit as he is not in a position to immediately apply the newly acquired knowledge. The manager is actually looking for recipes. Should a book contain lots of jargon and (worse still) mathematical formulae, he is likely to close the book after reading the first five pages, and donate it as a present to a staff member. Managers just want fast and simple answers to their problems (Mintzberg 1996). They seek tools to perform a task or to do something in a different manner. But the task in question, that which needs to be done, is in itself actually more important (Drucker 1994). The manager usually directs and takes decisions based on experience. This is an interesting observation, because most managers are gradually convinced that "on the job" training is not the fastest way. It is the case that the manager often gets by on his basic tertiary education, which, in general, did not even touch upon the problems of leadership and organisation. Decision-making based on experience is OK under relatively stable conditions. However, now that technology and society are changing at an increasingly faster pace, this is becoming more difficult. Science is also developing in this area. The problem, on the one hand, is how to convey this science to the hasty manager who still believes that the ability to "organise" is something that either comes naturally or not, and on the other hand, to identify what the manager should choose as a further area for study.

Ad 3. There is a fashion-conscious aspect to business management science. Gurus launch new techniques with much ado. These are often just basically good ideas for particular aspects of business practice, but they are promoted and sold as "the" solution to all organisational problems. ISO-9000, total quality management (TQM), logistics, lean production, business process redesign (BPR), workflow management, Six Sigma, etc., are the buzzwords of the last few years. Each of these techniques focuses however on just one aspect of the whole picture.

An *ISO-9000* quality certificate is currently a "must" for companies and it is now penetrating service and administrative organisations. This certificate is not only concerned with improving and controlling the quality of the processes but also with guaranteeing the future results of these processes. The intention of ISO-9000 is that one scrutinises the complete organisation. In practice, this has actually degenerated to merely recording the existing procedures. When gaps exist, they are filled. It is not about creating a better organisation but acquiring a certificate. More and more customers demand this. In determining these procedures, one elaborates on the procedures and procedure flow schemes from scientific management. This leads to, at the most, an improvement in the existing situation, but it does not foster innovation. For example, consider an officially certified procedure wherein, at a particular step, it is stated that employee X should examine the form and initial it. What he should look out for is not stated; the reason for initialling it is unclear—perhaps to confirm that the employee has seen it? (And what is the use, or function of this?). Or that he agrees; in which case, which authorities and standards are valid here? This array of worthless procedural descriptions was never the intention of ISO-9000, but it is what it has been reduced to. It has been wryly observed that one can easily manufacture concrete life jackets without losing the quality certificate. The year 2000 saw the introduction of a new, improved version of ISO-9000. This means a leap forward. Now, attention must be paid to controlling the process and to customer satisfaction and a management audit needs to be performed at frequent intervals.

Total quality management (TQM) is concerned with the control of quality. Statistical process control is a key aspect of this approach. But TQM goes further than this. The production process is considered one integrated system in which each successive department or employee is the "customer" of the previous one in the process. Each customer, whether internal or external, must be kept happy. TQM demands organisational change and a greater delegation of power. Cost cutting can be achieved through improving quality. TQM is more than just implementing quality circles in each department, as it views the organisation as a chain of individual, independent processes, with customer satisfaction being the ultimate goal.

TQM concentrates on one aspect, i.e. quality, and assumes that if this is improved all other aspects of the process will also be improved. TQM has often failed. Quality improvement and radical changes in the organisational structure to improve efficiency do not appear to go well together. TQM views the organisation as a system that is primarily concerned with satisfying customers. TQM also assumes that, in the long term, no conflicting interests will exist between employees and the organisation and between the organisation and its shareholders. This is, in our opinion, an unrealistic starting point. We see that as markets collapse, management abandons TQM and goes "back to basics".

Logistics and *supply chain management* are concerned with the flow of materials or of orders through the company: the shortening of throughput time; the reduction of stocks in hand; the simplification of the flow by reducing the number of intermediate steps between companies and departments. The goal is to improve the reliability of the delivery time and reduce the costs. Logistics often ignore the product design and unquestioningly accept the structure of the existing organisation whilst, for example, a different product design and a production structure with groups may realise even greater reductions in the production time. Applying logistics in a functional organisation structure is merely toying in the margin. Furthermore, logistics often limits itself to the areas of manufacture and distribution and leaves the initial trajectory, from product conception to manufacture, out of the picture.

Lean production is particularly applicable to mass production and focuses on a reduction of the throughput time, stock and costs.

Business process redesign (BPR) attempts to shorten throughput time and achieve an improvement in the control of processes and lower costs through the analysis and simplification of the transformation processes. In this technique derived from informatics, it is noticeable that little or no attention is paid to the circuits needed to control the processes. BPR can lead to great improvements, but it is often such that BPR has little influence on the end results of companies. The critique of BPR is growing and an improvement of its theoretical support is urgently required (Prakken 1995) (ten Bos 1997).

Workflow management is primarily an aid for visualising and analysing the operational process as a model. The question of whether this to-be-streamlined process is actually necessary at all in an organisation is not asked. Also missing is the intentional design of the necessary circuits for the control of quantity and quality. It was only as recently as 1997 that we saw the careful entry of the control paradigm in this area (Verhoef and Joosten 1996; van de Berg and Kusters 1997).

These approaches all originate from process thinking. This is why they have caused an actual breakthrough in "thinking in processes" as advocated by us. No publication on such methods, however, has provided a clear underlying theory. It is therefore impossible to arrive at an integration of the different trends or to show their exact limitations. However, the theory in this book forms the platform upon which we can acknowledge and supplement the shortcomings of the methods. This theory is also largely incorporated in sociotechnology.

Ad 4. Much of the scientific interest has been in problems that are not perceived by the manager as being the "most important" at the time. Science should precede practice, but it is an open question as to whether science has ignored an important field. Today's managers are struggling with problems resulting from the increasing complexity and changing attitudes of individuals and society. The complexity of the relationship between a variety of elements and phenomena of the organisational structures that are required to help understand that increasing complexity is problematic. How do we analyse such complex problems? How do we arrive at organisation structures that can withstand present and future problems? It is not about optimising existing processes but about new concepts. It is primarily about retaining or improving the effectiveness and flexibility of the organisation and far less about improving efficiency. The business analysts should certainly not lose sight of this analysis of complexity and integration. Also, product design is becoming increasingly complex and multi-disciplinary.

1.3 Conceptual Approach

This book closes the gap between theory and practice from three different viewpoints:

1. Extending the multidisciplinary character as far as possible
2. To find a solution, we should start with a common perception of the problem
3. Combining qualitative and quantitative modelling.

Ad 1. Each discipline or domain created its own systems approach with its own terminology. This has become a major problem, especially for complex system design (automated or not), where many disciplines come together. Currently all disciplines use some kind of a "system" concept to deal with complex problems. For example, business science considers organisations to be *combined social, technical and economical systems*; logistics emphasizes an integrated approach to dealing with an *operational system*; information technology developed several approaches to the design of *information systems*. They all construct models to formulate problems and find solutions. However, significant differences are evident between the models of each discipline. This creates a big problem for the project manager who should be able to keep all designs tuned into the system's goals. Moreover, each discipline tries to extend its modelling capacities further when confronted with limitations. For example, the Object Oriented Change and Learning modelling language is used to design and understand business systems as a whole, but is completely based on information technology (Swanstrom 1998).

This book presents a systems approach which is abstract and conceptual, but that contains all of that is required to elaborate a design further for each discipline involved. It views the company as an integrated whole and can place into perspective and combine the contributions from the many other disciplines. In this way, it supports collaboration in a design project and provides a way to save the principal decisions and assumptions during a design project in a systematic way.

Ad 2. In addition, education is mainly solution-oriented. From the very start, students are only trained to solve well-defined problems. In practice, however, they will be confronted with situations where the problem is not as well-defined as it was at school. Mostly they translate the situation into a problem that they recognize, with the consequence that although they solve that problem correctly, it does not appear to be the correct problem.

There is a clear need for people who are able to analyse a problematic situation and unravel it into isolated and clear problems.

The method used here supports the *analysis* of problems as well as the *design* of a solution.

Ad 3. Finally, there is a clear gap between qualitative modelling and quantitative modelling. Up until 1960, scientific research in the area of business management was aimed at improving the tools and techniques that emanated from scientific management, a highly quantitative approach. Towards the end of the 1960s the disadvantages of scientific management were becoming more and more obvious. The tools of scientific management were primarily focused on individual tasks and their optimisation. One presumed that the sum of the optimised work places would result in an optimally functioning company. In practice, it is now clear that in order to achieve an overall optimal situation we need to make many of the component parts sub-optimal. For example, we need to introduce overcapacity in a certain area (sub-optimal) to achieve the shortest possible throughput time in all other areas of the business. Much scientific analysis has since been performed in this area. It is the systems approach that primarily focuses first on the whole and then on the component parts; it thinks in terms of processes and process functions, and in doing so provides a better insight into the flows through the company. However, that approach has become a specialisation too, supported by the growth in the number of management and staff levels over the last decade. A "system" consists of a structure and shows behaviour, but the systems approach in business science mainly focuses on the structure. The quantitative modelling approaches should contribute to the dimensions of the structure, but until now there has been no clear modelling approach that translates a structural model into a behavioural model. In the field of simulation, for example, an appropriate modelling theory does not exist, although Zeigler mentions the process interaction approach (Zeigler et al. 2000).

This book not only presents an approach to defining *structure models*, but also provides a direct translation of these models into a *behavioural concept* that is ready to be simulated.

Systems theory and the systems approach must be primarily seen as a thought framework for personal use. It is possible to effectively pass on our findings to others without using systems jargon. It is not about whether our own organisation is "ready" for the application of the systems approach. It is an aid for "unique, *individual*" thinking. The manager needs a greater insight into what needs to happen within his department, which, in turn, interacts with the other surrounding departments. The problems mount up and many managers work day and night to remain in control. If the systems approach assists in permitting these managers to reduce the lengths of their actual working days, it is performing a very useful function. In our experience, thinking in terms of systems can partly assist the manager. It is not, however, a miracle cure. It offers a certain systematic method of thinking about problems. It provides better insight and transparency. It is a tool that can lead to a higher level of abstraction of contemplation of concrete situations. Faster conceptualisation is facilitated. We work less intuitively and can better place our experiences. Discussions with representatives of other disciplines become clearer. Multi-disciplinary working is facilitated.

1.4 The Structure of this Book

This book consists of three parts. The first part, Chaps. 2–7, covers the basic concept for modelling a system structure and behaviour in terms of processes and control circuits. The concepts of structure and time-dependent "behaviour" are combined in order to complete system modelling. Part I principally describes a fundamental approach to analysing industrial systems that emphasizes a concept that can be used by all of the disciplines involved and creates a logical systematic combination of quantitative and qualitative modelling. This approach is used for the analysis of industrial systems. Part II, Chaps. 8–10, is concerned with the use of these models in the design of (future) systems. Finally, Part III, Chaps. 3, 6 and 11, contains three comprehensive cases, which are drawn partly from our own practical experiences.

Our approach to writing this book has been to directly illustrate each theoretical concept with a practical example. However, being able to understand and absorb the knowledge provided in this book is not the same as being able to apply such knowledge to a real situation. The ability to apply knowledge is only gained by personal practice and not by studying examples and solutions as provided by others. The structure of the cases and exercises is such that demands are continually being placed on the reader's personal discipline. Before reading the solution, the reader is first expected to spend at least a quarter of an hour mulling over the problem and arriving at independent solutions to certain case-oriented questions. It is only through such practice that the essence of the material presented becomes clear, putting the reader in a position to apply the knowledge gained to his specific problems.

Both the theoretical parts and the application exercises have been exhaustively tested on students majoring in the discipline of Production Engineering and Logistics in the Faculty of Mechanical, Maritime and Materials Engineering of Delft University of Technology, and several Master Programmes in Business Management. In these Master Programmes, the students hail from every possible discipline and business branch: technology, informatics, sociology, law, economics, psychology and medicine. They are university and polytechnic graduates as well as employees from industry, government, banks, health services and other service sector employees. No discipline is faced with insurmountable problems.

References

van de Berg M, Kusters R (1997) Workflow management en systemen (Workflow management and systems). Tijdschrift Informatie, March 1997, pp 27–37

ten Bos R (1997) Business Process Redesign. Bedrijfskunde 1:56–66

Drucker P (1994) The theory of the business. Harvard Bus Rev, Sept 1994, pp 95–104

Mintzberg H (1996) Musings on management. Harvard Bus Rev, July 1996, p 61

Prakken B (1995) Business process reengineering: kan het (nog) beter (Business process reengineering: can it (still) get better?). Tijdschrift Bedrijfskunde 4:73–77

Swanstrom E (1998) Creating organizations with the OOCL method. Wiley, Chichester, UK

Verhoef D, Joosten S (1996) Een conceptueel raamwerk voor workflow management (A conceptual framework for workflow management). Informatie, Dec 1996, pp 51–58

Zeigler BP, Praehofer H, Kim TG (2000) Theory of modeling and simulation. Academic, New York

Chapter 2
Systems Concepts

Abstract. First of all, we will define and explain the most important, basic concepts from the systems approach. We readily admit that this is the most tedious chapter but knowledge of these concepts and their mutual interaction is absolutely imperative to be able to apply the systems approach and to think in terms of systems and processes in practical problems. The concepts of "system", "element" and "relation" will be sharply defined, together with the concepts derived from these definitions. In order to describe a system completely, the concepts of "structure" and "behaviour" are required. Furthermore, subsystems and aspectsystems are distinguished; this distinction is required for the correct modelling of systems. We will also define the key items of our approach: "function" and "task". Finally, the system boundary will be discussed.

2.1 System

The word "system" is a very general term. It is derived from the Greek verb meaning "to compile". The literature harbours dozens of definitions, as follows:

- A system is a collection of mathematical relations whereby a composition of physical objects is described
- A system is a man-made "whole" of interactive factors, variables (Lievegoed 1993)
- A system is a purposeful, ordered, coherent "whole" of related items and their constituent parts (dictionary; Koenen-Endepols)
- All that is not chaos is a system (Boulding).

The problem with all these definitions is that they all contain inherent limitations. The first definition confines itself to mathematical relations and to physical objects. A composition of non-physical objects is therefore not a system. In addition, the relations must be determined using mathematical formulae. The second definition requires human intervention in order to be able to speak of a system. Systems formed by nature are therefore not considered systems. The third definition demands that an arrangement must exist and that this arrangement must contribute to a certain goal. Groups of related parts, for which we cannot determine the goal, are thus not regarded as systems.

All of these definitions have two essential features in common, namely:

1. A collection of elements
2. Interaction between the elements.

We can discern the group of interacting elements within a greater whole. In the total reality, there are even more elements, elements which we do not acknowledge when studying the discerned group of elements. These latter elements can however be related to the elements in the total reality. We can thus *discern* a system within the total reality but we cannot *dissociate* the system from the total enveloping reality. Should we dissociate the group of related elements from the total reality, we run the risk of severing important relations with this reality. The group of elements we choose to study as a system within the total reality depends on the research requirements or on the researcher's own interests. The term "system" is therefore a way of thinking, *a way of looking at things*, that is dependent on the purpose for which we intend to use it. The definition must allow for this.

Definition

A *system* is, depending on the researcher's goal, a collection of elements that is discernable within the total reality. These discernable elements have mutual relationships and (eventually) relationships with other elements from the total reality. This definition introduces concepts that require further explanation.

Elements (Objects, Components, Entities)

These are the smallest parts considered by a researcher in view of his goals.

For this particular problem, they are not considered to be composed of even smaller building blocks. The sociologist, who studies a company as a system, will regard the company's people and machinery as elements. The company doctor, concerned with research into the effects of dust on the incidence of sick leave, will regard the dust particles, the lungs and, where necessary, other organs in the human body as elements. The engineer, charged with designing a new machine for a company, will regard the machine parts as elements. The material scientist, charged with advising on the choice of materials for these parts, will view atoms as elements of his system.

In mathematical systems thinking, it is common to refer to "objects". This is, in our opinion, too limiting. In many systems, not just *ob*jects but also *sub*jects, inert and living elements respectively, play a role. That is why we use the term element here, which can account for both. In addition, elements can be both material and non-material. With material elements such as the components of a machine or human organs, the system is *concrete*. Here, "concrete" means that it actually exists, is tangible and can be observed. The opposite of concrete is abstract. "Abstract" means separated from the material; intangible. Systems also exist where the elements are concepts; for example, capacity and resistance. These concepts have a mutual relationship that can sometimes be expressed in formulae. The related conceptual apparatus in itself can also be seen as a system. These are *abstract systems*, just like systems of services, of natural numbers in a numbering system or of feelings in a system of feelings. The concepts in this book also form an interrelated whole and are, as such, also an abstract system.

Content

We refer to the sum of the collection of elements as the content of the system. This is directly comparable to a "parts list" of a drawing that provides an overview of all of the parts that will appear in the drawing.

Attributes

The elements have certain properties. These can be physical, geometric, aesthetic, social, etc. An individual taken as an element has properties such as length, a face, and a character.

Seen qualitatively, an attribute often has different facets. The face can feel rough to the touch or look sympathetic, etc. The size is also one of the facets, and size has a certain *value*: the length of the individual is 1.85 m, and the face is very friendly.

Relationships

Relationships exist between elements. These relationships denote a particular interaction between the elements. In an abstract system, these are conceptual interactions. In a concrete system there is dynamic exchange. The elements influence each other. These influences can be mutual or one-sided. But what is the nature of this influence? This means that the characteristics of one element can change the values of the characteristics of another element and eventually vice versa. Characteristics that a particular element did not possess in the first place (had a zero value) can obtain a value other than zero under the influence of another element. We can say that the initial missing element is cultured. This distinction, that a property can have a zero value, can be important in gaining a clear insight. When rearing children, we try to reduce or suppress those characteristics we consider to be rated undesirable by society and preferably assign them a zero value, whilst encouraging the expression of or teaching other desirable characteristics, and increase their value through, for example, education. The term "relation" can also imply the positioning of the elements with respect to each other.

Relationships also have characteristics, but we will not delve deeper into this area here. Much has been published on this topic, particularly in the discipline of sociology.

Structure

The enumeration of the collection of relationships is referred to as the structure of the system. The parts list of the drawing provides the content; the actual drawing provides information on the structure, such as place and form relationships.

Universe

Here we refer to the total reality, i.e. all elements and relationships, known and unknown, in reality.

According to the definition, a system is a group of elements that the observer distinguishes within that universe. The elements of the system have interrelationships but can, according to the definition, also have relationships with other elements in the universe. Not all elements in the total reality will have relationships with the elements of the system. We thus distinguish the meaning of "environment".

Environment

The environment belonging to the system under consideration is comprises those elements from the universe that influence the characteristics, or the value of the characteristics, of the system's elements; or in reverse, are influenced by the system. When we consider a company as a system, the elements of the prevailing society form a system of higher order that influences the company and the elements within. Society therefore forms a definite part of the environment of the company's system. On the other hand, the planet Saturn, which is also an element of the universe, probably does not influence the company's system and therefore does not belong to its environment. This is really unclear for a particular company on another continent. This can or cannot influence the company. The actual components of the system's environment are often difficult to determine.

In principle, the system's environment is the collection of objects and subjects in the universe that influence the elements of the system but are not constituents of the system. The environment is part of the universe.

We therefore make a distinction within the definition of "structure". All inter-element relationships within the system form the *internal* structure. All relationships with elements from the environment form the *external* structure.

Emergence

Emergence is the principle that whole entities (groups, elements) display characteristics that are only meaningful when they are assigned to the whole and cannot be reduced to the individual elements. For example, the odour of ammonia or the image that appears as we progress with the piecing together of a jigsaw puzzle. Each model of a system of human activity displays, as a complete entity, characteristics that emanate from the activities of the system's elements and its structure, but which are not retraceable to these. These are *emergent characteristics* of the whole (Checkland and Scholes 1990; Hitchins 1992).

Summary

Universe (the total reality)

- Environment (the elements of the universe that have relationships with elements of the system)

System

- Concrete (tangible)
  - Content (summing up of the elements)
  - Structure (summing up of the relationships)
    - Internal (the inter-element relationships within the system)
    - External (the relationships between some elements within the system with elements from the environment)
- Abstract (intangible)
  - Content
  - Structure
    - Internal
    - External

2.2 Subsystems and Aspectsystems

In order to obtain a clearer insight into a complex system, it has been shown to be extremely useful to differentiate the system into subsystems and aspectsystems. (De Leeuw 2000).

A system is composed of elements and relationships. We can therefore use two methods to differentiate partial systems: subsystems and aspectsystems.

2.2.1 Subsystem

Definition

A *subsystem* is a partial collection of the elements in the system whereby all the original relationships between these elements remain unchanged.

We therefore think of a division into subsystems as a division into groups of elements whereby all the original relationships between the elements of such a group, and their relationships with the other system elements, retain their original properties.

A subsystem completely conforms to our definition of a system. A subsystem is a system whereby the original system forms an important part of the environment of the now differentiated subsystem; for example the starting motor of a car engine. The motor can be considered to be a subsystem of the car. The car forms the environment of the motor. Nevertheless, at one stage lower the starting motor forms a subsystem of the motor whereby the motor and part of the car form the starting motor's environment. In technical systems, we usually differentiate subsystems as those groups of elements that collectively assist or aid in the greater system. In a company's system or organisation, the division between what we want to differentiate as a system or subsystem and what is the environment is often less obvious than in technical systems. Depending on the problem definition, it is often recommended that those subsystems of an organisation should be chosen that form a more or less independent part or that fulfil a certain process function in the whole.

2.2.2 Aspectsystem

Definition

An *aspectsystem* is a partial collection of the relationships in the system whereby all the original elements remain unchanged.

The relationships within an aspectsystem are generally of a singular type. The aspect we wish to examine determines the type of relationships that we distinguish in the partial collection.

The remaining relationships are not considered here. As such, we could separate the following aspectsystems in a motor:

- The thermodynamic aspectsystem, such as the conversion of chemical energy into kinetic energy, resulting in heat transfer and material expansion
- The kinematics aspectsystem: the predicted movements that the parts must make with respect to each other
- The tribology aspectsystem: the mutual friction of the moving parts and the lubrication required

- The spatial aspectsystem: the positioning of the compulsory parts with respect to each other
- The control aspectsystem: the controlled progression of the process
- The strength aspectsystem, such that the parts can tolerate the apparent forces in action
- The maintenance aspectsystem: the approachability, the ability to replace and the life expectancy of the different parts.

In a company, it is possible to distinguish the following aspectsystems:

- The technological aspectsystem, e.g. the rows of machinery in a production line
- The economical aspectsystem, i.e. the cash flows or the value added flows
- The social aspectsystem, i.e. the relationships between the workers and between the workers and the machinery
- The spatial aspectsystem, i.e. the physical positioning of the departments within the factory walls or of people and machinery within a department
- The information aspectsystem, i.e. the flow of the planning and control data between the departments or the employees
- The political aspectsystem, e.g. where is the balance of power and the political influence with respect to decision-making?

The above-mentioned relationships and more can be found in both a motor and a company. Each type of relationship can be studied as an aspectsystem.

Relationships also exist between the individual aspectsystems. Such relationships are termed *inter-relationships*. Our knowledge of these latter relationships is limited. We often omit the word "*aspect*". We usually only describe certain aspects of a system, certain relationships between the elements. This omission presents a real danger. In the long term, we forget that we have excluded, without reason, various other relationships that are present. We base our conclusions and decisions only on that which we have studied by that one aspectsystem. This can have disastrous consequences: for example, buying a digitally controlled milling machine but neglecting to instruct one of the workers (element) in its use (relationship) or forgetting to adjust the work preparations. Here, one has defined a systems limit that is too narrow, namely, only around the machine itself.

Sometimes an aspectsystem is also referred to as a partial system (De Leeuw 2000). In mathematical, partial differentiation, one assumes that the remaining relationships are constant. In an aspectsystem, we go further and completely exclude the remaining relationships. The danger of such action has already been pointed out. In order to avoid any confusion with respect to this point, the term "aspectsystem" will continue to be used. It is perhaps unnecessary for us to point out that the term "social aspectsystem" does not have the same meaning as the term "social system" as used in sociology. A social system in sociology is a system where some or all of the elements are people. Also, in this respect, we cannot omit the word "aspect". In this book, the term "social aspectsystem" refers to all relationships between people who belong to the system we are studying.

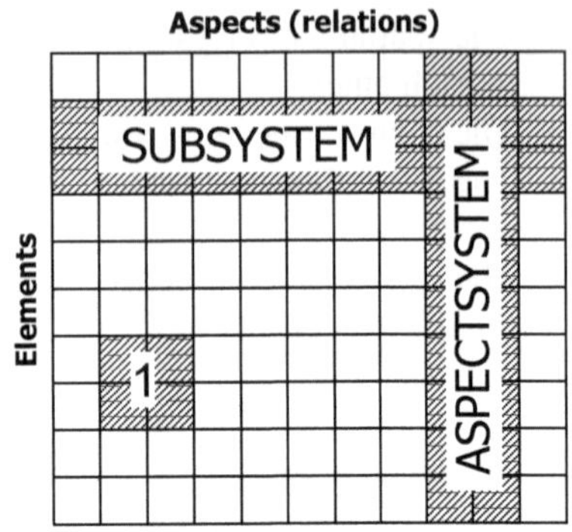

Fig. 2.1 Subsystems and aspectsystems

Summary

In subsystems, we distinguish groups of elements whilst retaining all the relationships between the elements.

In aspectsystems, we distinguish relationships whilst retaining all the elements. Both approaches can occur simultaneously such that we view some aspects of a subsystem (1 in Fig. 2.1), an aspect-subsystem, or sub-aspectsystem. The matrix in Fig. 2.1 depicts this.

In his attempts to understand the universe, man mainly studies subsystems within aspectsystems. The number of relationships between all elements that together form the universe is, after all, unimaginably large for the limited capacity of the human mind.

So:

System

- Subsystem (a group of elements retaining all relationships)
  Aspect-subsystem (a group of elements within which we only look at certain functions)
- Aspectsystem (certain relationships retaining all elements)
  Sub-aspectsystem (certain relationships, but now only within a group of elements).

Examples

Take a group of students in a lecture theatre as a system. The system's boundary lies at the walls of the theatre. If we consider all of the relationships between the students in the left half of the theatre only, then we are looking at a subsystem. On the other hand, if we consider only certain relationships between the students in the entire theatre—for example, their positions with respect to each other and their eventual memberships of a student society—then we are studying an aspectsystem. Now, if for the whole theatre we draw the system's boundary around those students that are members, and then we get one or more aspect-subsystems. If we take only the left half of the theatre and look at their memberships, then we have a sub-aspectsystem.

Consider another example. If we consider a building a system, then the subsystems are, among others, a floor, a room and a stairwell. The aesthetic appearance of the building is an aspectsystem just as much as the colours. This is the architect's domain. The civil engineer looks at other aspectsystems such as weight, strength and measurements. The business specialist will look at the transport aspectsystem, among others. The communications technician looks at the telephone aspectsystem, a part of the communication aspectsystem.

2.3 State, Process and Behaviour

Each of the elements in a system has certain characteristics. These characteristics have certain values. The elements can mutually influence each other such that one or more characteristics change in value. We then speak of relationships between the elements.

2.3.1 State

We can, for now, define the state of a system at a defined time as the value of the properties at that time in the system.

An *event* occurs when the value of the property of an element changes; that is, when the state of a system changes. When one event inevitably leads to another event, we talk of an *activity*. An activity takes time. The state of a system at a particular moment is a result of previous events. The definition of state thus includes the *memory* of the system. All relevant historical information is stored here. To describe the state, we must provide an overview of the value of the properties of all of the elements at that time.

Sometimes not just the values of the properties but also the relationships between the elements can change over time. In this case, there is talk of a *changing structure*. With an *unchanging structure*, we must be aware that a structure can *appear* not to change. When the relationships continuously change with a definite cycle-time and we continue to take measurements at a determined point in that cycle-time, then the structure appears to be unchanged to the observer. Due to the chosen interval between measurements, the system appears to have an unchanged structure.

It can even be the case that a number of elements are removed from or replaced in the system. This changes the content and thus the structure because a number of relationships also change. In such cases we must not only provide a description of the value of the properties but also of the content and structure at that time.

Managers are primarily concerned with systems that must fulfil a function in their environment; that is, make a contribution to a greater whole. Some functions can be fulfilled by unchanging *static* systems, such as a suspension bridge or a pillar of the bridge, or a system of roads. Within such a system we find elements and relationships but no events. At least, that is not the intention. To fulfil other functions, events and activities must take place within the system. Those are time-dependent systems in which processes occur.

Beware: the terms *static* and *time-dependent systems* refer respectively to the absence or presence of a process in the system. Changing or unchanging refers to the structure of the system, for example:

- Car motor: unchanging structure, time-dependent system
- Company: changing structure, time-dependent system
- Map: unchanging structure, static system (i.e. not a process)
- Stamp collection: changing structure, static system.

Time-dependent systems often need various supplies from the environment, such as energy, materials, and ideas, etc., in order to fulfil their specific functions. Some systems, such as companies, fulfil their function in the environment by delivering products or services as required by the environment. Generally, in such systems we can differentiate between:

- Input
- Throughput
- Output.

The simplest depiction of such a system to which we will repeatedly refer is displayed in Fig. 2.2.

This image is based on the flow of matter.

The flowing elements may or may not have their own identities, e.g. cars on a production line or oil in a refinery. It is essential to define exactly what it is that flows. In the case of the car factory, we can regard the raw materials that must be transformed into cars as flowing elements just like the received orders that are converted to fulfilled orders for cars. In the first case, we are concerned with raw material flow, and in the second case with the "information material" flow. The matter that flows is indicated next to the process arrows. So our simplest scheme of a car factory with respect to the materials would be represented as in Fig. 2.3.

In a system where a process occurs, we can distinguish between the *permanent* elements and the *temporary* elements. Only for the latter can we talk of a flow

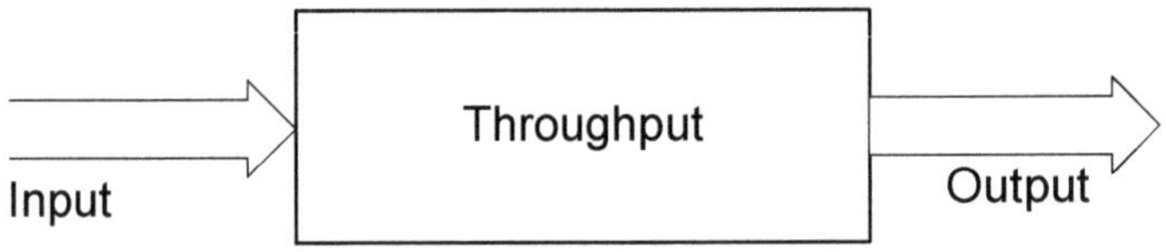

Fig. 2.2 Simplest scheme of a time-dependent system (based on the flow of matter)

Fig. 2.3 Simplest scheme of a car factory

time. These temporary elements are continuously supplied to the system where they are transformed by various activities during the throughput to become the output demanded by the environment, plus eventually waste. These activities collectively form the process. The permanent elements fulfil functions in that process.

2.3.2 Process

Definition

A *process* is a series of transformations that occur during throughput which result in a change of the input elements in place, position, form, size, function, property or any other characteristic.

The way in which the activities are performed, their respective sequence, etc., is determined by the properties of the permanent and temporary elements and by the relationships that exist among the permanent elements and with the temporary elements.

Content, structure and process are thus inextricably bound together by an indissoluble tie and must therefore be studied together. This book will primarily examine industrial systems, systems with processes that are clearly focused on a particular goal. The activities are individually linked by information flows, the information that is required for the execution, at the right time, right place, right manner and in the right combination of the right activities. The activities in such a process are mostly performed by subsystems, each of which has its own determined function in the process and makes its own contribution to the process. Through that process, the system fulfils its function in its environment. The fulfilment of that function in the environment is the system's *goal*. Each element and each subsystem delivers its own contribution in the process towards realising the system's goals. In a well-organised, technical system, no element can be removed without reducing the total function fulfilment of the system. A car without wheels or without petrol cannot fulfil its function.

Doubling the number of elements such that when one is removed the other takes its place (redundancy) is sometimes necessary to reduce risk. The second element is therefore not superfluous because the goal of the system is to also reliably fulfil the systems function. When a certain element stops functioning in a living system (in vivo), this function is sometimes taken over by other elements.

2.3.3 Behaviour

A time-dependent system will display certain behaviour within the period in which the system is being studied. Therefore, the researcher must take the time period into consideration when studying the system. To what extent does the researcher wish to make future predictions about the system? What is his time-span? We denote this with the term *phase system* (De Leeuw 2000).

There are thus three possible subdivisions of a system:

- *Subsystem* is a group of elements, e.g. a department
- *Aspectsystem* is a group of relationships, e.g. the interpersonal relationships in a company
- *Phase* system, the time span within which a system is observed, e.g. within a period of one month.

At the very start of cybernetics, RosenBlueth et al. (1943) defined "behaviour" as "any change of an entity with respect to its surroundings". They declared this behaviouristic approach to be the examination of its output and the relation of this output to the input. They contrast this to the "fundamental approach", in which the structure, properties, and intrinsic organization of the object are studied, rather than the relationship between the object and its environment. The term "any change of" must be clarified with respect to that "relationship".

To achieve this, the definition of in 't Veld (2002) of "any change" is considered. He defines the behaviour of a system as "the way in which the system reacts to internal and external conditions, to certain inputs and their transformation". He relates behaviour to the "state" of the system. "The state at some point in time of a system with a given structure is a set of values that together with the input signal at that point in time unambiguously determine the output signal". He considers the set of values the result of input signals from the past. A value must belong to a "property" of the system.

The term "unambiguously" complicates the interpretation of this definition. Apparently the state of the system is not fully described by the values of its internal properties but must also include the values of external properties, which may influence the internal properties (by means of input signals). The conclusion is that the state of a system must reflect both the internal and external conditions. Consequently, there are two sources by which a property value may change: the system itself or an external source (another system or the environment).

A change of the state is therefore a consequence of a change in the set of property values or a change in the input signals.

We define behaviour in the following way:

Behaviour is the property of an element that describes the way in which the state of the element together with its input results in output.

According to this definition, an element has two major properties: state and behaviour. The behaviour cannot be expressed by a simple value, but must somehow express the way in which the state will change.

When we can completely predict the output from a known input, then the behaviour is *completely determined* (deterministic). When chance plays a role and the output is only predictable with a measured probability, we refer to the behaviour as being *stochastic.* A system is in a *steady state* when it displays behaviour that is completely determined and repeatable in time, whereby the behaviour in one interval is similar to the behaviour in another interval. In the case of stochastic behaviour, this means that the probabilities must have a determined value.

Mathematically, we could suggest that the probability distribution of the possible states is constant. This is the case, for example, for an electric clock. However, the behaviour of a system can also change with time: so-called passing or *transient* behaviour. The probability distribution of the possible states in that case is not constant. A system that displays signs of growth or decline, expansion or reduction has transient behaviour and is often referred to as a transient system. Such a system often displays one-off reactions that cannot be repeated; for example a child's first steps and his/her reactions to this event.

2.4 Goal, Function and Task

In the previous text the terms goal, function and task were mentioned in passing. A clear distinction between insight into and the coherence of these terms is essential for thinking in terms of processes and systems. Many people have problems when they go to apply these terms to a concrete situation.

Function and task are often confused in common parlance. When we want to design or analyse systems it is imperative to make a clear distinction between these two terms. Malotaux (1997) distinguishes this by making a distinction between:

Task	Function
What the element does	Its purpose
The actual work	The (unintentional) effect of it in the greater whole

Malotaux defines the terms "function" and "task" as follows.

Definition

The *function* of an element (object or subject) is that which is brought about by that element towards satisfying a need of the greater whole. In short: the desired contribution of a part to a greater whole of which it is a constituent.

Definition

The *task* (usually tasks) is concerned with what needs to happen or needs to be done in order that the contribution is realised such that the function is fulfilled.

The task is concerned with the actual work, the activities. The function is concerned with the context of the task within the greater whole.

With function, it is the net result on the environment that counts and not *how* the system accomplishes this. It is *not* about the activity itself.

Both terms are in fact extensions of each other, just like the throughput and the output. The task is concerned with the activities that take place within the element or subsystem. The function is concerned with the consequences of these activities in the environment of the subsystem. We think of functions as contributions to a greater whole. As such, we are currently not concerned with the manner by which the subsystem realises this internally. By thinking in this way we keep all our options open for alternative ways of realising this contribution.

Here is a rule of thumb: it is a function when the same contribution can be realised by different means. For example: the *function* to be fulfilled is "producing an electrical current". A bicycle dynamo can achieve this but so too can batteries. It is thus a function. The *task* of the bicycle dynamo is "to convert the wheel motion into electrical current", and this is only possible with the bicycle dynamo. The task of the battery is: "to produce an electrical current by means of a chemical reaction". A function is almost always a verb. A noun is usually a resource. We must *fulfil functions* whereas tasks are *implemented.*

We shall therefore often design a system by first determining the functions that must be fulfilled in the system to realise the system's goal. In this analytical phase, it is essential to clearly distinguish the functions to be fulfilled, irrespective of whether two functions can be fulfilled by one and the same organ. For example, the wall of a house has different functions, such as partition (visual separation of the inside of the house from the environment, heat insulation, sound insulation) and transmission (supporting the floor above or the roof). Normally, all of these functions are assigned to the one organ, "the wall". By assigning these different functions to separate organs, we can arrive at surprising new constructions. In particular, when using new materials, this approach can be essential for maximising the usefulness of the new materials. Another example: a hospital's goal may be formulated as follows: "The healing of patients and/or the easing of their suffering and/or patient reassurance with respect to their perceived suffering".

This is the function that the hospital must fulfil in society. In general, fulfilling the function in the environment is the goal of the system. To achieve this, various functions must be fulfilled within the hospital, such as diagnosis, treatment, nursing, caring. The question of whether these are functions rather than tasks often arises. The answer to this is again dependent on the question on which the attention is focused. For example, we can view "nursing" in two ways: as an activity in itself (task) or as a contribution to the healing process (function). The function is paired with the desired result. These functions in the hospital can be assigned to separate departments (subsystems). In turn, it is the aim of, for example, the X-ray department to fulfil a function in the hospital, namely to provide the opportunity to perform an internal examination of a patient without resorting to an operation.

Therefore, the tasks of the X-ray department are to take X-rays but also to keep abreast of the technical developments that can contribute to new and better ways of fulfilling that function. The hospital system therefore needs certain functions. To achieve this it creates subsystems that are charged with fulfilling these functions in the hospital system. This means that a variety of tasks (activities) must take place within the subsystem. Seen from the perspective of the total hospital system, subsystems are merely means to realise the hospital's goals.

Summary

The goal of such a goal-oriented system is to fulfil certain functions in its environment, functions that the environment needs for its processes. Within that system, consecutive processes must occur. In these processes, various process functions are fulfilled for which tasks must be performed. The functions are assigned to certain subsystems. The goal of that subsystem is to fulfil the required function in the system's process. The subsystem becomes a *means* to achieve the system's goal.

Figure 2.4 illustrates our concept of a function. By transforming input into output, *requirements* are being fulfilled to some extent, which is expressed by the *performance*. Requirement and performance are directly related to the goal of the function.

A function is less time-dependent than a task. In the aforementioned X-ray department, many other means and methods have become available in addition to the X-ray taking facility, such as ultrasonic diagnostics. Whilst the function remains the same, the tasks change, and it may be that the name of the department also needs to be changed. Within a time-independent function, we see a continuous shift in task implementation from man to machine, which is brought about by rapidly advancing technology. In offices, we witness a rapid shift from man to the computer, where the function, the contribution to the greater whole, remains unchanged. It is not just because the function is afforded a longer life than the task to be performed that we increasingly take the perspective of functions instead of tasks in the design of technical processes and in organisations, but also because,

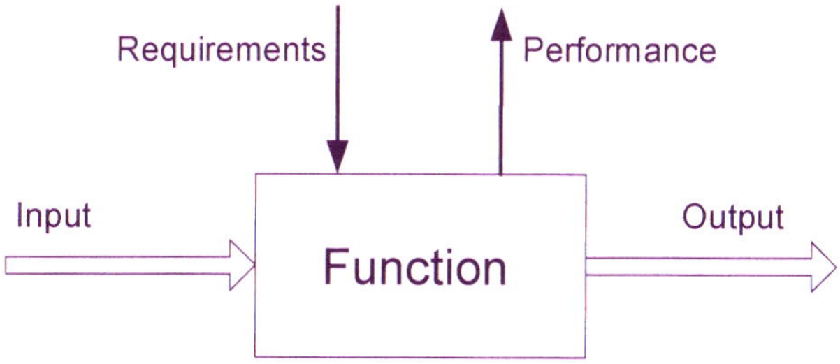

Fig. 2.4 Simplest scheme of a function

when designing functions, various alternatives remain open in terms of the manner with which (the task) a function will be fulfilled. We then determine which functions are needed within the process and which relationships must be realised. We can, in fact, speak of the design of the functional structure. The process functions form the elements: the relationships between these functions form the structure. In organisation science, the term "functional structure" has had a completely different meaning for some ninety years. To avoid confusion, this book will stick to the classical meaning of the term "functional structure", as derived from organisational science. The former is indicated as "*process function system*" or "system of process functions". In such a system, the relationships between the functions are primarily determined by the process cohesion. Functions change in time more slowly than tasks. In the long run, it is possible that the need for a certain function in the environment can change or even completely disappear. As such, the goal of the system or subsystem that fulfilled this function changes. Should that system wish to continue to exist then it must focus itself on a new goal.

Question

As practice material for the previous issues addressed, indicate which elements are not correctly represented in the "bakery" system displayed in Fig. 2.5. Why is this and how should it have been incorporated into the scheme?

Answer

In the scheme displayed in Fig. 2.5 only "prepare dough" and "inspect" are functions. The other elements are not functions.

"Flour" is not a function; it is a flowing element. It does not belong in a process function block but should be placed next to the process arrows. The same applies to "cookies".

"Oven" is not a function but a means of fulfilling the function. The function is "baking".

The "customer" is also the means of performing the function, which is "consuming".

The functions "preparing dough" and "baking" are often collectively placed in one *department*, the "bakery".

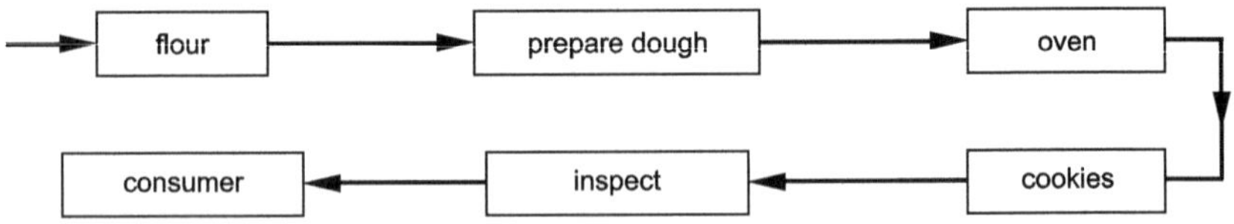

Fig. 2.5 Question: system of process functions

2.5 System and Environment

The system's environment is composed of those elements from the total reality that exert influence on or are influenced by the elements of the system. We now distinguish between open and closed systems. An *open system* is a system whose elements interact with elements from the environment. A *closed system* knows no interaction with the environment; in other words, the system's environment is considered to be empty. This distinction also depends on the manner of consideration. If, whilst studying an open system, we accept for simplicity's sake that no relationships exist with the environment, then we are actually considering the open system to be a closed system for the purpose of the study.

It is also possible that confusion can arise here because, in electricity and hydraulics, the term open circuit implies one where the process flow or the information flow is impeded by an "open" switch or closed tap. When we flip the switch closed then the process takes place and the circuit is closed. In systems thinking, the difference between open and closed systems is completely different. Better terms would be respectively: not isolated and isolated systems.

All systems with an input and output are difficult to study as closed systems. We tend to ignore too many essential relationships. A system that does not have input during the test period but does have output is called a *free system*. An open system has input and output.

2.5.1 The System Boundary

A system is a cohesive whole of elements. In order to distinguish a system from its environment, we need to draw a *system boundary*. The eventual input and output traverse this boundary. In principle, this boundary can be chosen freely. In practice it is determined by the method of consideration chosen by the researcher as being the most adequate for his particular question. We must be careful not to draw this boundary too narrowly. This could lead to the possibility that the cause of the problem as presented falls outside of the chosen system and thus evades detection. If we chose a royal boundary, the whole becomes more complex and unmanageable. Choosing the boundary is a difficult issue. All things are connected and if we are not careful, we may conclude that in order to solve a specific problem in the maintenance department of a firm, we must first change the complete universe. Inexperienced researchers are often inclined to draw a royal boundary. In contrast, experienced researchers often display the inclination to draw too narrow boundaries; Company myopia can be an important reason for this behaviour.

How we define the system to be studied is dependent on the goals that we set the system and the questions to be answered. The (artificial) system boundary is therefore also dependent on both the ability to measure the goals and the various means to achieve these goals.

A system should form a cohesive whole. In practice, we can therefore only consider a group of elements as a system when their mutual cohesion is greater than their cohesion with the elements in the environment. In order to be sufficiently manageable, the system boundary should be chosen as follows:

- Such that the exchange of the *temporary* elements—such as material or patients or, with further abstraction, data—across the boundary is less than within the boundary. In this case, it should be that the exchange takes place via a number (as small as possible) of permanent elements.
- And/or on the basis of the number of relationships dissected by the boundary. Whether or not we wish to consider a particular element as belonging to the system is thus dependent on whether the number of relationships among the elements already incorporated in the system is greater or less than the number of relationships with the environment. It is not about all conceivable relationships here, only relationships that are considered relevant to the problem.
- And/or on the basis of the energy required for transmission through the boundary. Transmission through the boundary requires more energy than a transmission that takes place within or beyond the boundary.
- And/or such that we are able to clearly formulate the function of the chosen system in the environment. Does the (sub)system have clear emergent properties? The omission or addition of an element can influence this.

The system's boundary is primarily determined by the goal of the study. Within this parameter, we must attempt to choose the system's boundary such that the least number of relationships are severed and the least possible number of temporary elements crosses the boundary.

The latter must, in addition, occur via the least possible number of permanent elements. Always choose the system boundary to occur at a discontinuity in the process.

We are often presented with a number of possibilities in choosing this boundary. Confusion can occur if we choose the boundary that is, unnecessarily, on the royal side; choosing too narrow a boundary means we risk overlooking important aspects. The choice of boundary is often an iterative process.

We usually base the choice on practical considerations. A couple of examples are as follows:

- When we wish to study a family, we can place the boundary around the constituent elements, the family members. We could also include the home help in the system or consider the actual house as an element in the system.
- The strain on an electrical or a mechanical system can be considered either as part of the system or as an output. Some car manufacturers give the engine power in so-called DIN-kW, others in SAE-kW. When measuring the performance in SAE-kW, all aids, such as the dynamo, are removed. When using the DIN measurement, all accessories are present, as in the case of the car. The performance in SAE-kW is therefore always greater than that in DIN-kW. It can be useful to be aware of this difference when comparing cars.

- When researching certain phenomena in a dishwasher we may or may not regard the operator as part of the system, just as we may the dirty crockery.
- When investigating the actual organisation of an X-ray department, we could place the systems boundary at the walls of the department, but we could also include the operating theatre or other departments in the system.

Whether we have chosen meaningful boundaries for the purpose of solving the problem becomes apparent when the results obtained are confronted with reality. It is primarily about making a purposeful choice for the boundary in view of the problem to be analysed.

2.6 Some Other Definitions

In Sect. 2.3, *steady state* is defined as the state of a system that materialises when the behaviour of the system is repeatable in time and when the behaviour in one period is similar to the behaviour in another period. The opposite of this is *transient*, passing behaviour. In a closed system, the eventual steady-state equilibrium situation is only determined by the initial conditions. In an open system, an eventual equilibrium situation is actually maintained through a continuous exchange of temporary elements between the system and the environment. It is a dynamic state of equilibrium. *Homeostasis* is the term used to define the overall equilibrium maintained by a living organism. Such an open system can attain time-independent equilibrium independent of the initial conditions. We call this the principal of *equifinality*; that is, the maintenance of a steadfast goal-orientedness based on a dynamic equilibrium under changing conditions.

Finally, the concepts of *wholeness* versus *independence* continue to be important. For the latter, one no longer refers to a system but instead to an *aggregate*.

If each element of a system has relationships with all of the other elements in the system, then a change in the value of a property of a random element causes a change in all of the other elements and thus in the whole system, and so that system is referred to as being a *wholeness*. The coherence of all elements in the system is reflected in the collection of relationships, and therefore in the structure. The extent of coherence is a property of the structure of the system. This is called the *coherence grade.*

Summary

Depending on the goals set by the researcher, a system is a distinguishable collection of mutually related elements within the total reality. Within the system, a distinction has been made between subsystems and aspectsystems. The following definitions were discussed: state of the system, processes within that system and system behaviour. The difference between task and function was reviewed.

Emphasis was placed on the point that in the systems approach one should think in terms of "functions" being the constituent elements of the system, not the "tasks" or the "means". Finally, the problem of determining the system boundary was discussed.

References

Checkland P, Scholes J (1990) Soft systems methodology in action. Wiley, Chichester, UK

Hitchins DK (1992) Putting systems to work. Wiley, Chichester, UK

de Leeuw ACJ (2000) Bedrijfskundig management (Business management). Van Gorcum, Assan

Lievegoed BC (1993) Organisatie in ontwikkeling (Organisation in development). Lemniscaat BV, Rotterdam

Rosenblueth A, Wiener N, Bigelow J (1943) Behaviour, purpose and teleology. Philos Sci 10(1):18–24

in 't Veld J (2002) Analyse van organisatieproblemen, een toepassing van denken in systemen en processen (Analysis of organisation problems, an application of thinking in terms of systems and processes). Stenfert Kroese, Leiden

Chapter 3
Case: The Flight Department

Abstract. The case in this chapter illustrates the use of the preceding theory, particularly the black box approach and both qualitative and quantitative reasoning. It is a case that was analysed in real life (except for the simulation part). The data are the real data present at that time. However, a series of extra complications have been skipped. Readers that are interested in learning more about this case will find a complete description in (in 't Veld and Muller 1966). We selected this case from the practical experiences of in 't Veld as an example, because the actual results of the analysis can be shown for a period of nearly three years. At the time, the complete analysis took about three months. The part that is offered here as an exercise took from two to three weeks. If the preceding theory is applied, however, this can be shortened significantly, but this theory was unknown at the time. No technical knowledge is required to solve this case. Although the example presents a situation from the 1960s, it still is very relevant to our times in terms of testing other types of airplanes.

3.1 Case History

The Fokker aircraft factory receives, together with some other companies, an order to construct 350 units of the F104-G Starfighter. The G indicates that this is the European model, which contains far more electronics than the other models. It is (or was at that time) a highly advanced aircraft. It is modern mechanically, but aligns closely with the existing operational and technological experience of Fokker. However, the electronics are completely new. Even in the United States only two prototypes are flying with these electronics. The customers ask for the delivery of one airplane every 1.5 working days. With 240 working days in a year, this means 160 airplanes a year.

H.P.M. Veeke, J.A. Ottjes, G. Lodewijks, *The Delft Systems Approach*,

3.2 Problem Description

Production starts in the first year. In February of the second year the production line has reached its required production rate of one aircraft every 1.5 days. On March 1 of that year, 70 airplanes were delivered to the flight department and every 1.5 days one more airplane is added. During their stay in the flight department, the airplanes are tested in the air by test pilots. Any shortcoming that is detected is repaired by the flight department. When a test flight shows no more failures, the airplane is marked "off-test", and after a final check the airplane can be delivered to the customer.

Although the production line had already produced 70 airplanes, only 20 airplanes were delivered to the customer on March 1.

Figure 3.1 shows the currents until that date. The upper line represents the delivery from the production line to the flight department. The production rate is one every 1.5 working days, which means 13 1/3 airplanes each month. The lower line

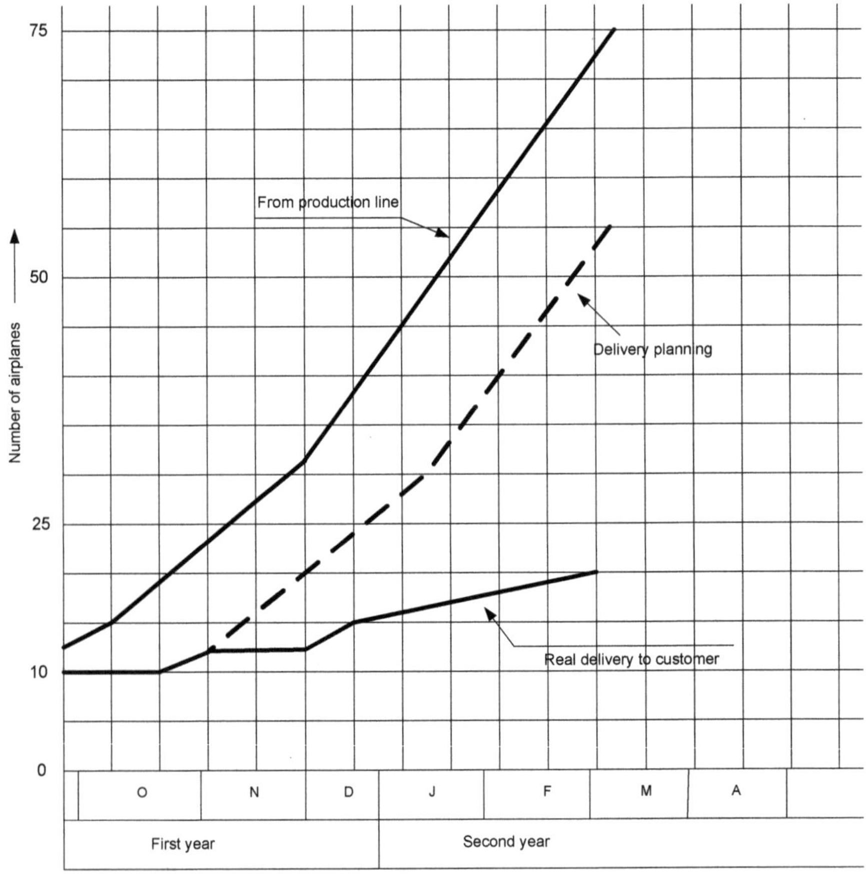

Fig. 3.1 Production planning and progress until March 1

shows the real delivery rate to the customer. The dotted line in-between represents the planning for this delivery. On March 1 the delivery runs 31 airplanes slow with respect to the planning. In the flight department there are about 50 airplanes at that moment. Due to lack of hall space (there is only room in the hall for 12 planes), they are parked on the terrain, which is sometimes covered with snow.

In former contracts, a lead time of six weeks was included for the flight department. It was assumed that the weather was always good to fly, which means that test flights are possible every day. This clause is also included in the F104 contract. The horizontal distance between the lines for delivery from the production line and the planned delivery to the customer is therefore six weeks in Fig. 3.1. Now, 100% good weather will never occur in The Netherlands. To maintain the registration for each airplane in order to be able to claim the days on which testing was impossible as acceptable delays is a tremendous job for those 50 airplanes, and so the subsequent processing of these data results in a pile of paperwork.

The delivery rate of the production line in Fig. 3.1 is indeed one airplane every 1.5 days. But the trend of the delivery line to the customer is not more than one airplane every nine days. The customer demands an immediate acceleration of the delivery.

The questions asked of management were: what do you need in terms of people, space and test equipment? Which organisational approach are you going to apply? Show these data to the customer in an acceptable way.

3.3 Problem Analysis

An analysis of the available data is started, in the hope of producing insight. The data that are collected are based on a five-day working week:

- The production line delivers one airplane every 1.5 days.
- Three days are required to prepare each airplane for the first test flight.
- The total number of flights until "off-test" ("off-test" means that all systems are functioning well during a flight and the airplane satisfies all requirements) varies from three to fifteen flights; on average nine flights are needed before off-test.
- Each flight takes about 1.5 hours, but pilots are not always available immediately or an immediate test flight is impossible for other reasons. Therefore one should take one day of throughput time into account for each flight.
- After each flight except for the last one, the deficiencies need to be solved. The throughput time for repair varies from one to nine days and does not depend on how many flights have already been done. The average throughput time for solving deficiencies appears to be 3.7 days; see Fig. 3.2 for the frequency distribution. At this stage, it is estimated—after lengthy consideration—that an average of three days will be feasible at short notice. This will be assumed from now on.

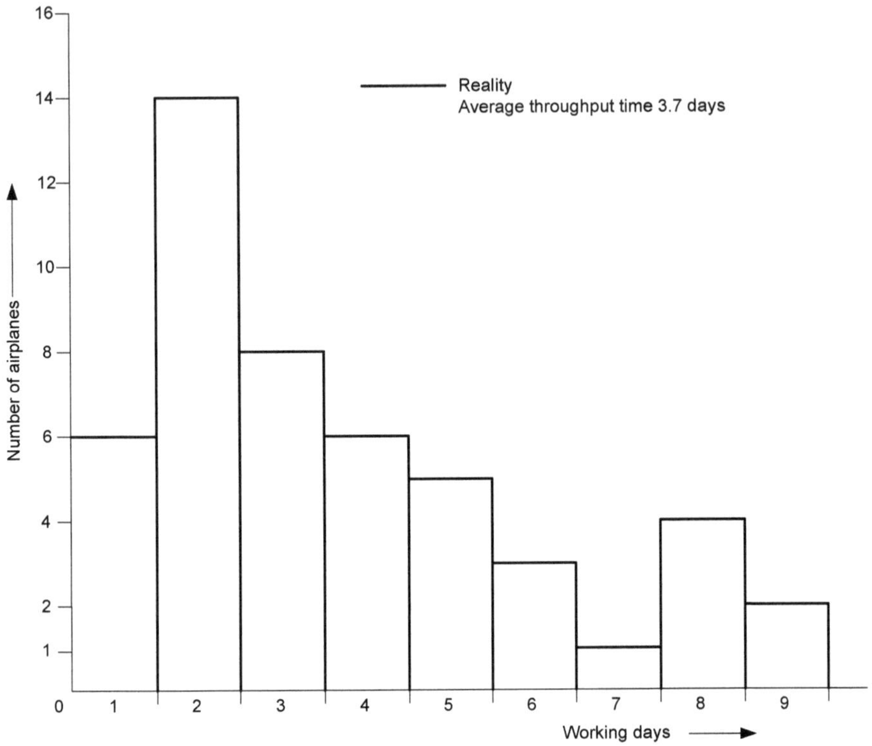

Fig. 3.2 Frequencies of throughput times for solving deficiencies after a test flight

- Sometimes an airplane expires before starting a test flight: a so-called abort. This may happen before a random test flight number. Some aborts can be solved at the starting location and the test flight can still be performed within a few hours. Sometimes it takes more time to solve the abort: a so-called prolonged abort. In such a case the plane should return to the hall. Solving such a prolonged abort takes just as long as normal deficiencies after a flight (3.7 days). In these cases, one also counts on a throughput time of three days in the future. Prolonged aborts occur in 10% of all starts. An abort is not considered a test flight and is not contained in the abovementioned average number of flights.
- After reaching "off-test", the airplane must be prepared at the flight department for a delivery check; this takes one day. The actual delivery check takes place at the production line by the military inspectors present there. "Ready for delivery check" is the contractual point that is coupled to the clause of six weeks of unbroken good weather.

With these data, a model can be constructed of the process that takes place at the flight department. Of course, we should define the goal of the model: which problem do we wish to solve with the model? Different problems require different models. The reader is now expected to think the problem over for about 20 minutes and to answer the following questions. The goal of the model is expressed by the questions.

Questions

A1 Develop a process model for the path of the airplanes through the flight department. Base it on the normal course during the normal rate of the production line of 13 1/3 airplanes each month. Leave the (apparent) arrears out of your consideration.

Using this model, calculate:

A2 The average number of positions in the hall of the flight department required for each of the different activities

A3 The average number of test flights that should be performed in order to realise the program each day

A4 The average number of airplanes that should be delivered by the group that solves deficiencies each day

A5 The average total number of airplanes that should be circulating in the flight department

A6 The average total throughput time for the flight department assuming 100% good weather

A7 What are your conclusions with respect to the contract and the available space in the hall?

Remark
Of course you can include parts of days in your calculations, but not half or quarter flights. A quarter of a flight is a normal flight, because the duration of a flight is not mentioned in these data.

Answers to Questions A1–A7

A1 *Develop a process model*
Approaching the flight department as a black box, we can only state that one airplane enters each 1.5 days and that an airplane should leave at the same rate. See Fig. 3.3.

This model is very simple, but we cannot do very much with it. The available data cannot be placed in it very clearly. Therefore we zoom deeper into one aggregation layer. This is shown in Fig. 3.4.

This already offers a better overview, but the black box for flying and solving deficiencies is still unclear. Therefore we have to zoom in at that location one aggregation layer further down; see Fig. 3.5.

Fig. 3.3 The flight department as a black box

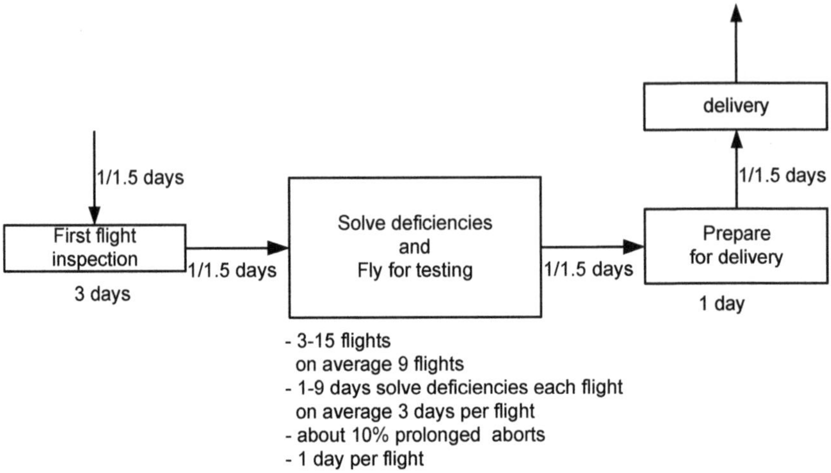

Fig. 3.4 First simple model

Each airplane passes through the loop "fly and solve deficiencies" on average nine times, plus one more for a prolonged abort (that is 10% of the starts). This can be drawn as shown in Fig. 3.3.

In Fig. 3.6 the assumption is made that the abort occurs between the third and fourth flights. Exactly when the abort occurs is not important; it is only important to know that each airplane suffers on average one prolonged abort, which will result in 10% aborts on average. Note however that the abort cannot be drawn as the last one, because after the reasons for an abort have been solved another test flight is required.

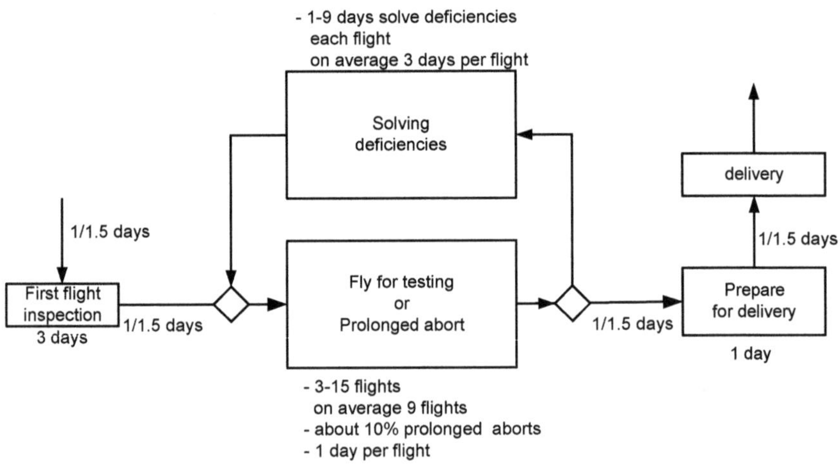

Fig. 3.5 Second simple model

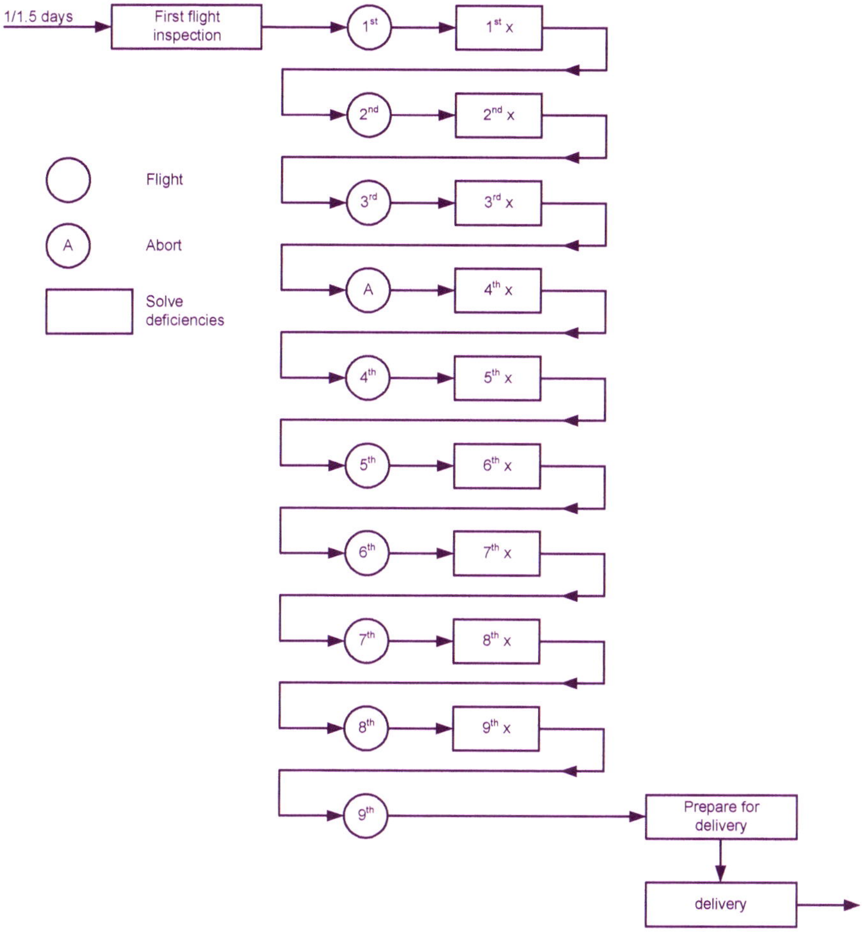

Fig. 3.6 Course of an airplane through the flight department

Figure 3.6 is another representation of Fig. 3.5 that is presented here because it is easy to make a particular mistake. For nine test flights and one abort, deficiencies must be solved only nine times. The last test flight has no deficiencies. This can also be seen in Fig. 3.5.

A2 *Using this black box model, Fig. 3.5, calculate the average number of positions required in the halls*

The inspection for the first test flight takes three days. The input flow rate is on average one airplane every 1.5 days; in other words, 2/3 of an airplane each day. A throughput time of three days gives: 2/3 airplane per day × 3 days of throughput time = 2 airplanes in progress. So two positions are required. If the input flow rate was once every three days then 3/3 = one position would have been required.

Questions

1. We change the numbers. The throughput time through the black box "first flight inspection" is four days and the input rate is three airplanes per day. How many positions are required inside the black box?
2. The throughput time through the black box remains four days, but the input rate is changed to one airplane each 0.5 days. How many positions are required then?

Answers

In the first case, twelve positions are required and in the second case eight positions are needed.

The calculations above are based on Little's formula (Little 1961), which is:

$$N = \lambda \times D$$

Where: N = the average number of elements in a system (read black box)
λ = the average input flow rate
D = the average throughput time through the system.

This formula holds for all processes and is always applicable. The significance of this formula is often underestimated. If during a design process one needs to estimate how many resources are required for a process, the formula gives a first estimate of the dimensions. If it is applied when analysing a process for problem solving (as it is here), it provides an easy estimate of the minimum resources required to achieve the required performance. If there are more resources in reality than required according this formula, one can easily calculate the occupation of the resources.

Returning to the case of the flight department, we now calculate the number of positions required to solve deficiencies. The throughput time is on average three days. If it is only necessary to solve deficiencies once for each plane, then $2/3 \times 3$ = two positions are required. As already shown, it is necessary to solve deficiencies an average of nine times for each plane, eight times after a flight, and once after an abort. So, in order to solve deficiencies, 9×2 positions = 18 positions are required.

The preparation for delivery takes one day and requires 2/3 of a position, or in reality a position that is free for half a day after a day of occupation.

So, in total, the number of positions required is:

first flight inspection	2 positions
solving deficiencies	18 positions
prepare for delivery	1 position
Total	21 positions

A3 *What is the average number of test flights required each working day in order to realise the program?*

Every airplane should make an average of nine flights. The delivery asks for one airplane each 1.5 working days, so 2/3 of an airplane a day. Therefore, the required number of test flights each working day is:

$2/3 \times 9 = 6$ flights a day

This includes the first flights.

A4 *What is the average number of airplanes that should be processed by the group that solves deficiencies each day?*

On average, deficiencies must be solved for each airplane nine times, and one airplane should flow through the model every 1.5 days, so the black box "solving deficiencies" should deliver an average of $2/3 \times 9$ = six airplanes each day. The 10% of aborts are already included. This can also be calculated by noting that according to A2 18 positions are required. So 18 positions / 3 days of throughput time = 6 airplanes each day.

A5 *What is the average number of airplanes that should circulate in the flight department?*

In the black box "first flight inspection"	2	airplanes
In the black box "test flights and abort"	6 2/3	airplanes
In the black box "solving deficiencies"	18	airplanes
In the black box "prepare for delivery"	2/3	airplanes
Total	27 1/3	airplanes

A6 *What is the total average throughput time assuming 100% good weather?*

In the black box "first flight inspection"	=	3 days
In the black box "test flights and abort"	10×1 =	10 days
In the black box "solving deficiencies"	9×3 =	27 days
In the black box "prepare for delivery"	=	1 days
Total		41 days

We can check these results by putting flow rates next to the arrows in the model of Fig. 3.5. At the input and output side this has already been done, but not in the loop. At the first node of the loop (left side of Fig. 3.5), 2/3 of an airplane arrives each day from the "first flight inspection", as well as six airplanes a day from "solving deficiencies". The input flow rate into "fly for testing or prolonged abort" is therefore 6 2/3 airplanes each day. 2/3 of an airplane arrives off-test at the node after this black box and then moves straight on, while the remaining six airplanes move to "solving deficiencies".

The flow rate of six airplanes a day that passes into the black box should equal the output flow rate. So six airplanes leave the black box. With a throughput time of three days we need 18 positions. The calculations were correct.

Little's formula not only fits each black box of this model, but it also fits at the highest aggregation layer to the black box representing the whole flight department. The input flow rate there is also one airplane every 1.5 working days (see Fig. 3.3). We have already calculated that the throughput time through the whole Flight Department is 41 days. Little's formula leads then to $2/3 \times 41 = 27\ 1/3$ airplanes in the flight department on average. This agrees with our detailed calculation above.

A7 *What are your conclusions with respect to the contract and the available space in the hall?*

When the weather is always good enough to perform test flights, the throughput time is 41 days, which means more than eight weeks. Therefore, the contractual six weeks are not realistic and do not satisfy the real values of the variables. The flight department needs 21 positions in total, while the current hall only has room for 12 positions. The flight department should therefore be extended. After consultation with management and the customer, it was decided that the values of the variables presented were reasonable for the aircraft type F104-G and that the results of the previous calculations should be accepted.

3.4 Complications Due to Bad Weather

Until now it has been assumed that the weather is always suitable for flying. However, this does not reflect reality in The Netherlands. Sometimes the weather conditions make it impossible to fly on a particular day, and sometimes these conditions continue for more than a day. How do these weather conditions influence the model? Assuming that the number of positions calculated above are available and the average number of airplanes and resources are present at these positions, then what are the answers to the next set of questions?

Questions

B1 What are the consequences of one day of bad weather for the model of Fig. 3.5?

B2 What are the consequences of several (or more) consecutive bad weather days?

B3 What changes are required in the model to be able to take bad weather days into account?

B4 What delays will occur to the throughput time of an airplane in cases B1, B2 and B3? What delays will occur to the complete program of 350 airplanes in these three cases if the bad weather days occur according to the scheme below? In other words, how many days late will the delivery of the last airplane of the series be?
A study of the previous two years gave the following data for the periods of time for which it was impossible to fly and test an airplane like the F104-G:

Bad weather days in two years:

63 × 1 day	= 63 working days
25 × 2 consecutive days	= 50 days
11 × 3 consecutive days	= 33 days
7 × 4 consecutive days	= 28 days
3 × 5 consecutive days	= 15 days
1 × 6 consecutive days	= 6 days
1 × 7 consecutive days	= 7 days
Total	= 182 days

It is assumed the same scheme will hold for the next two years.

B5 What are the consequences of these bad weather days for the results calculated in A?

Remark
We have to use the assumption that no extra capacity can be obtained by doing overtime. The data are based on a situation in which the majority of the flight department works according a two-shift scheme. For these personnel it is legally forbidden to perform overtime.

You should take approximately 20 minutes to answer the questions.

Answers to Questions B1–B5

B1 *What are the consequences of one day of bad weather for the model of Fig. 3.5?*

The black box "solving deficiencies" delivers six airplanes every day, even on a bad weather day, when no test flights can take place. On these days, the black box should still be supplied with six airplanes in order to preserve the utilisation of 18 positions continuously. If no test flights are performed then these six airplanes will be not available. The six positions will stay unoccupied, and the people and resources on these positions run idle. If the six airplanes were to arrive one day later it would be impossible to make up arrears, because everything has been tuned to a rate of six a day and not more. Exactly the average required production capacity is available everywhere. If this capacity is not used for one day, the people will sit around and twiddle their thumbs, and that day is permanently lost.

One single bad weather day causes a permanent delay of one day for the whole program of 350 airplanes. All airplanes including the last one of the series will be delivered one day late. Actually, one stops the clock in the flight department for one day and then continues with the same rate. One is not able to speed up the rate, because the capacity is exactly tuned. However, the clock *outside* the Flight Department continues as usual. Another example: You drive your car at top speed and it takes you two hours to reach your destination. If you were to get a flat tyre, and the repair took 15 minutes, then you would definitely reach your destination after 2 hours and 15 minutes. Your car (the resource) cannot go faster than top speed. The delay of a quarter of an hour will have its effect until the destination.

B2 *What are the consequences of several (or more) consecutive bad weather days?*

On the second consecutive bad weather day, the hall will deliver six airplanes again. This leaves 12 positions idle, because there is no input yet. This day will also be permanently lost. After three consecutive days of bad weather the hall will become empty. All bad weather days are permanently lost.

B3 *What changes are required in the model to be able to take bad weather days into account?*

The airplanes that come from the black box "solving deficiencies" (from the hall) cannot be lined up for a test flight because there are still six airplanes waiting for good weather. The model should be extended with buffer stock in order to store these airplanes. Therefore, a "wait for testing" buffer is introduced into the model (see Fig. 3.7). So there are now 12 airplanes in total that were tested earlier and are waiting for good weather, comprising six airplanes waiting for "solving deficiencies" and six airplanes in buffer 1 that have just left the black box "solving

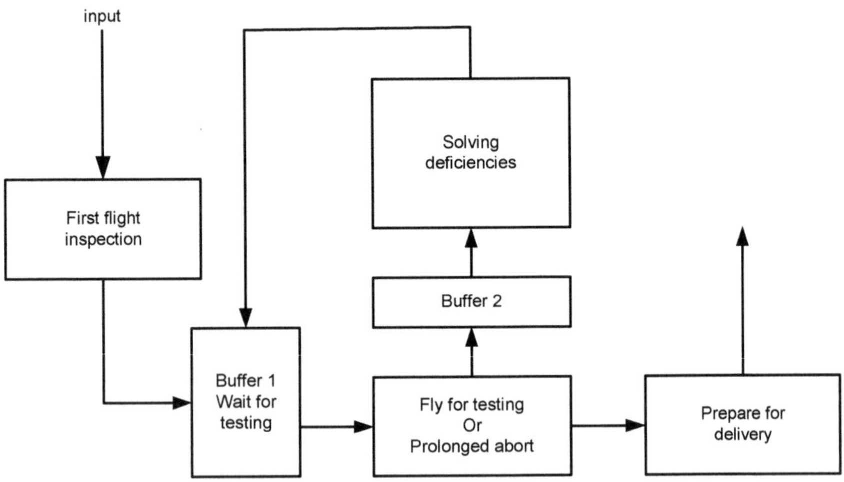

Fig. 3.7 Third simple model with buffers (compare with Fig. 3.5)

deficiencies"; plus the airplanes that have been delivered by the black box "first flight inspection". After three consecutive bad weather days there will be 18 airplanes waiting in buffer 1, six for "solving deficiencies", excluding the airplanes waiting for their first test flight, and the hall has become empty. This all happens fully autonomously. Nothing can be done about it.

However, the only reason for permanent delay is that the black box "solving deficiencies" has no input on a bad weather day. One could prevent a delay by keeping six airplanes in a reserve stock between "fly for testing" and "solving deficiencies" (Fig. 3.7). For this reason buffer 2 is introduced into the model. The consequence is that one should add six airplanes to the circulation in the flight department. Every airplane that has been flown will stay in that buffer for one day before it is fed into the black box "solving deficiencies". If one chooses to cover oneself against two or three bad weather days, one should add 12 or 18 airplanes to the buffer. The model now looks like Fig. 5.7.

If one covers oneself against three consecutive bad weather days by including 18 extra airplanes into the model, then these airplanes will be all in buffer 1 after three days of bad weather, while buffer 2 is still empty. If it becomes good weather indeed on the fourth day, then there will be 24 airplanes ready to take a test flight. The occupation of the pilots however is tuned to 6 test flights a day. Buffer 2 will stay empty, even if the weather stays good. Inclusion of extra airplanes into the model only has an effect the first time there are three consecutive bad weather days; after this a delay will occur after all during the next period of bad weather because buffer 2 is empty and stays empty. This can only be prevented by pushing airplanes from buffer 1 to buffer 2 at a faster rate than the average required rate. In reality the pilots are able to perform more than 6 test flights a day. Moreover, the model only includes the usual five working days per week. If the pilots would be prepared to perform overtime during the weekends (especially during the winter with the majority of bad weather days), then there would be a significant gain of capacity. Essentially we introduce extra capacity at one position of the model compared to the average required capacity. By doing this we are able to realise a higher flow rate. We select the simplest or cheapest position in the model. In practice this happened in this way.

B4 *Which delays occur in the throughput time of an airplane in cases B1, B2 and B3? Which delays will occur in the complete program of 350 airplanes in these three cases?*

The normal average throughput time was calculated in question A6. It is 41 days in good weather. Every bad weather day leads to a delay of one day. The total delay for the program of 350 airplanes equals the total number of bad weather days during the period of two years. That number equals 182 days, or 36 weeks. Under the condition of an extra buffer stock of six airplanes, the throughput time of one airplane in the model of Fig. 3.7 becomes 41 days + 9 × 1 day = 50 days. After every flight or abort the airplane should then wait 1 day in buffer 2 before it is processed by "solving deficiencies". If one covers oneself against three consecutive bad weather days by buffering 18 airplanes, the throughput time becomes 41 + 9 × 3 days = 68 days. Every airplane must then wait for three days in buffer 2 after each flight.

Covering for one bad weather day increases the throughput time by 9 × 1 days. This also holds for the program as a whole. The last airplane of the series will also be delivered nine days later. One bad weather day happens 63 times, but these do not cause further delays, and one day can also be subtracted from each period of bad weather that lasts longer than one day. The delay of the complete program will now become:

As a consequence of increase of throughput time	= 9 days
63 × 1 bad weather day	= 0
15 × 2 consecutive bad weather days	= 15
11 × 3 consecutive bad weather days	= 22
7 × 4 consecutive bad weather days	= 21
3 × 5 consecutive bad weather days	= 12
1 × 6 consecutive bad weather days	= 5
1 × 7 consecutive bad weather days	= 6
Total	= 90 days

Despite the increase in throughput time caused by the increased number of airplanes in circulation, the total delay will be 90 days instead of 182 days.

When one covers oneself against two consecutive bad weather days, it can be calculated analogously that the total delay will be only 61 days. The total delay will be only 47 days if three consecutive bad weather days are covered for. The throughput time would increase by 27 days (5.5 weeks) in this case.

If we buffer against four, five or six consecutive bad weather days, then the total delay to the program becomes 44, 48 or 55 days. In other words, the increase in the throughput time would dominate over bad weather when covering against five consecutive bad weather days. It was decided to cover the process against three consecutive days, because there is only a small difference in the total delays for buffers of three and four days, and because the reliability of the data is poor.

B5 *What are the consequences of these bad weather days for the results calculated in A?*

All calculations made in A stay valid; only the average throughput time increases, to 41 + 9 × 3 = 68 days (13.5 weeks), and the average number of airplanes in circulation becomes 27 1/3 + 6 × 3 = 45 1/3 airplanes.

It is clear from the previous calculations that the number of airplanes in circulation will increase if the output of the model stagnates for some reason. The production line however keeps delivering one airplane every 1.5 days. Because of this the throughput time will automatically increase. The process as a whole will be less well organised and it will be more difficult to compare it with the designed model. Therefore, it is decided to keep the number of airplanes in circulation constant. To do this, two "taps" are added to the model. Whenever an airplane passes

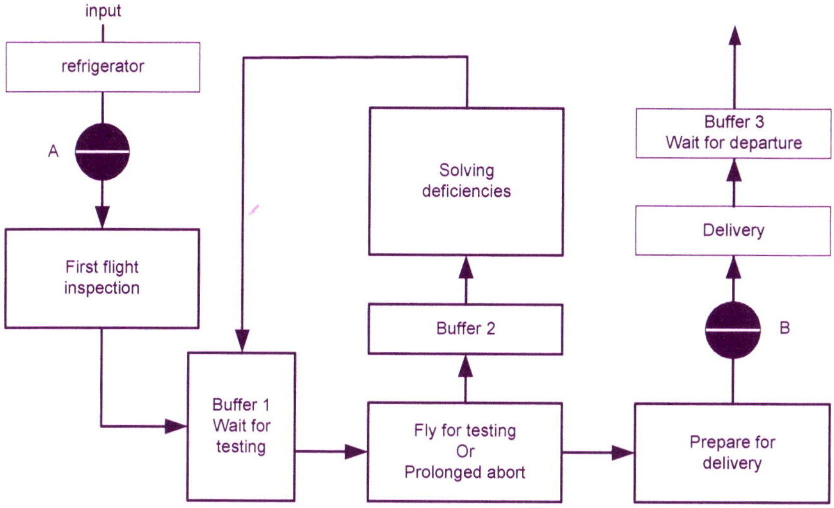

Fig. 3.8 Extended simple model with buffers

the output tap, one airplane is allowed to pass the input tap. So another buffer is required before the input tap, the so-called refrigerator.

Airplanes departing to the customer can also be troubled by bad weather. Therefore, a buffer is also needed at this stage. This buffer is positioned behind the output tap, because this tap represents the contractual point of "delivery". The model is now as shown in Fig. 3.8.

In order to continuously gauge whether reality still corresponds to the calculations in the model, the following variables should be regularly checked against tables and graphs. As assumed before, the input values are the standards that the variables should satisfy. These variables are:

- Input: one airplane every 1.5 days
- Average number of flights until off-test: nine
- Average output from "solving deficiencies": six
- % Prolonged aborts: 10% of the starts
- Total average number of airplanes in circulation between the taps A and B: 45 1/3 (and the refrigerator should be empty). In times of bad weather some airplanes may be waiting in the refrigerator
- Number of bad weather days to justify delays: if 47 days of delay, as calculated before, are accepted, other claims for delays will only arise when there are more than three consecutive bad weather days.

In order to compare reality continuously with these standards and eventually intervene when necessary, control loops are clearly required. However, they have not been drawn in the model, because the problem definition does not give rise to this.

Based on this study, the complete pile of paperwork for insurance claims for each airplane can be abolished. One should simply create and update a table containing the numbers of consecutive days when there was bad weather.

Section 3.2 stated that on March 1 there were about 50 airplanes in the flight department and that the delivery was 31 days behind schedule. The schedule was based on 100% good weather. The previous calculations showed that with a normal procedure and the existing bad weather distribution there should be about 45 airplanes in the flight department. This means an average throughput time of 45 × 1.5 days = 67.5 days = 13.5 weeks, which is not even close to the six weeks of consistently good weather. If we compare the actual number of 50 airplanes with the theoretically required number of 45, we can then conclude that we only appear to be behind by 31 airplanes. The delivery line of the planning is completely unrealistic and we would run far behind a realistic delivery line. The flight department should become overloaded according the model before the required delivery rate is reached. This argument is strengthened by the fact that the date of March 1 is the end of the winter period, which has many bad weather days. Equalizing the score is therefore out of the question.

The delivery rate did indeed start to rise after March 1 without any intervention in the workshop. That delivery acceleration would have occurred whether or not we performed this analysis. However, if we would have skipped the analysis we would not have understood what was really happening in the Flight Department, and we may have made expensive investments, like adding capacity. These investments would then have turned out to be superfluous, because the extra capacity could not be used.

3.5 Radar Complications

Our insight into the flight department was improved tremendously by developing the model and performing calculations based on it. The influences of different variables are reasonably well understood now. However, the resources required for each position is still unknown. This should clearly be kept as low as possible. In addition, we have calculated until now with averages, and the effects of dispersions in, for example, throughput time and when solving deficiencies (1–9 days with an average of three, see Fig. 3.2) are still an open question. In particular, radar is the cause of many complications. When solving radar deficiencies, special positions that allow the airplane to beam through a window in the wall are required, as well as test cars costing €250,000 a piece. Only a limited number of positions in the existing hall can be used for this purpose. This lack of radar test positions should be taken into account when constructing the new hall, as justified above.

Not every test flight will result in radar deficiencies. Expensive positions and test cars are not needed then. According to the test flight reports, 50% of the test flights show both radar and mechanical deficiencies and 50% only mechanical deficiencies. Radar deficiencies require considerably more time to solve than mechanical deficiencies. Figure 3.2 shows the frequency distribution of the throughput times for all airplanes. Figures 3.9 and 3.10 show the frequencies for airplanes with and without radar deficiencies separately.

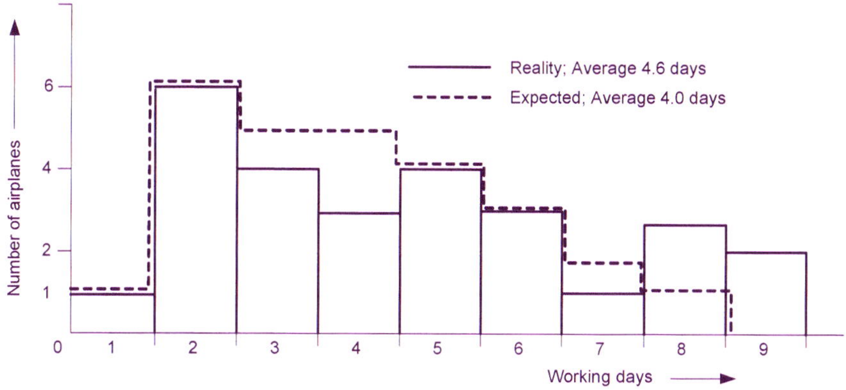

Fig. 3.9 Frequency distribution of time taken to solve the deficiencies of airplanes with radar deficiencies

The airplanes with radar deficiencies require a throughput time of 4.6 days on average (Fig. 5.9), while airplanes without radar deficiencies require on average 2.7 days. The average for all airplanes is of course still 3.7 days.

Previous calculations already assumed an improvement to 3.0 days on average. If we persevere in this assumption, we should assume an average throughput time of 4.0 days and 2.0 days for airplanes with both radar and mechanical deficiencies and airplanes with only mechanical deficiencies, respectively. This gives the frequency distributions shown with dotted lines in Figs. 3.9 and 3.10. While the total number of airplanes is the same for the distributions given by the dotted and solid lines, in each plot the dotted (expected) distribution is shifted a little to the left of the solid (actual) one. Further calculations use the frequency distributions given by the dotted lines.

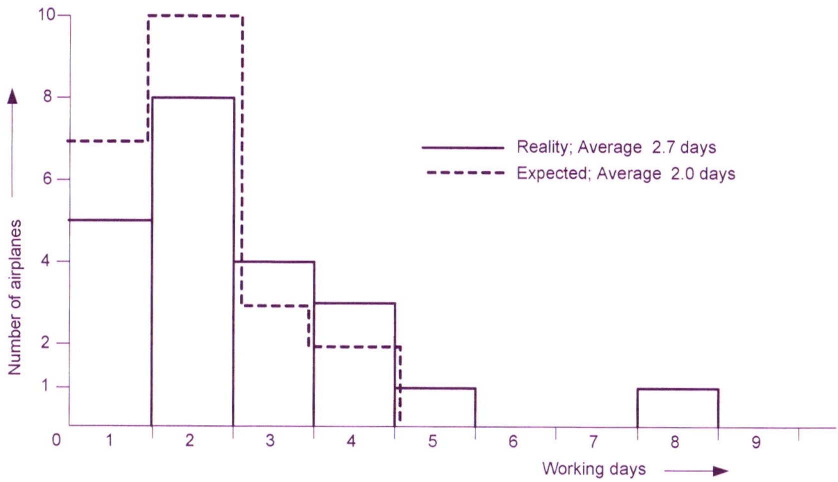

Fig. 3.10 Frequency distribution of time taken to solve the deficiencies of airplanes without radar deficiencies

Considering the price of €250,000 for a radar test car, it is also important to check whether this equipment is needed throughout the throughput time for an airplane with radar deficiencies. We find that this is *not* the case. The radar test car is only needed for some of the throughput time. The data for this are shown in the next table.

Table 3.1 Data on throughput time for airplanes with radar deficiencies

Total throughput time needed to solve deficiencies (days)	Of this, the throughput time needed to solve radar issues
1	1
2	1
3	2
4	2
5	2
6	3
7	3
8	4

For an airplane with radar deficiencies, it is decided that the mechanical deficiencies should be solved first, before the radar deficiencies are addressed. Otherwise there is a significant chance that the radar will break down again during the mechanical repair activities.

What are the consequences of these more detailed data on radar-related throughput times for the flight department's model shown in Fig. 3.7? There are three different ways to adapt the model.

Question

C1 *What are these possibilities, and which possibility is preferable? (Take about five minutes to think this over.)*

Answer to Question C1

The *first* possibility is to equip all of the positions with the facilities required to solve radar deficiencies. However, this solution is very expensive, because all positions will then need a window in the wall.

The *second* possibility is to equip special radar positions. The mechanical deficiencies are solved first, and then the airplane is transferred to a special radar position. This approach, however, works best when a special position is available whenever required. Therefore, a buffer would be required here, which would facilitate the independence of these types of activities. The model then looks like Fig. 3.11.

Fig. 3.11 A possible model for solving deficiencies for airplanes with and without radar deficiencies

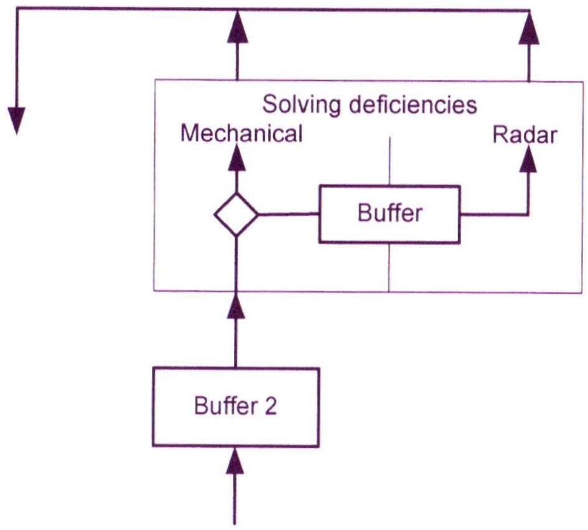

The third possibility is to fix all of the deficiencies of the airplanes with radar deficiencies completely at the radar positions, including the mechanical repair work. The airplanes without radar deficiencies would be prepared at the remaining positions. In this way, the black box "solving deficiencies" is split into two separate black boxes that can function completely independently of each other, and each needs its own buffer for this.

The last solution is preferable, because although the second solution requires less radar-equipped positions, the need for a buffer causes extra throughput time. The independence of both black boxes in the third solution makes the situation in the department easier to survey, and the area in question will be quieter. The model of the flight department can now be drawn as shown in Fig. 3.12.

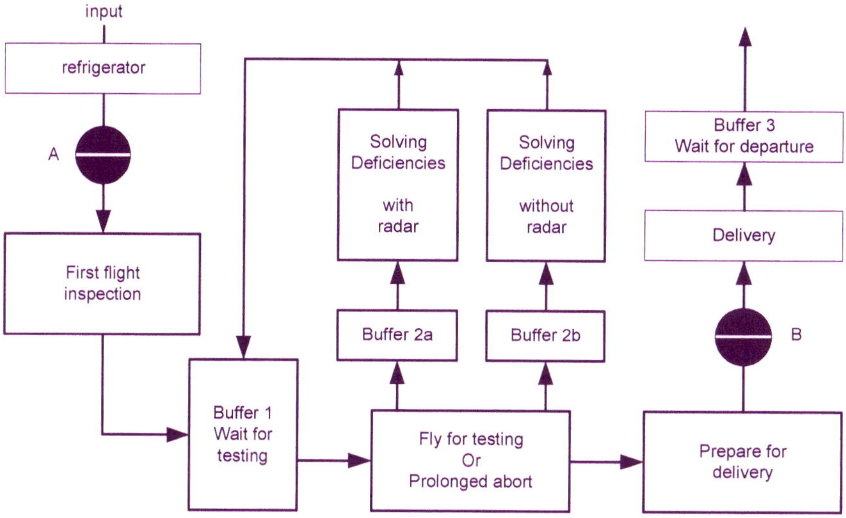

Fig. 3.12 The final model of the flight department

In Sect. 3.4 we calculated that deficiency solving requires 18 positions in order to reach the required rate of six airplanes a day. The distribution between both radar and mechanical deficiencies and mechanical deficiencies alone was found to be 50-50. Because of this 50-50 distribution, both black boxes should deliver the same number of airplanes each day: three each. However, the averages of the throughput times are 4.0 and 2.0 days respectively. So twice as many positions are required for airplanes with radar deficiencies than for airplanes with mechanical deficiencies only. Thus, 12 positions are needed for radar deficiencies and six positions for only mechanical deficiencies, a total of 18 positions.

As a result, as well as the variables mentioned previously, the following variables should be checked on a regular basis:

- Output of the black box "solving deficiencies with radar": three airplanes a day
- Output of the black box "solving deficiencies without radar": three airplanes a day
- The distribution between the two categories: 50-50.

Remark

The elements in the final model of Fig. 3.12 are process functions. It appeared to be very difficult to develop an operations research program from this model. Operations research thinks in terms of resources, the capacity of these resources and queues before the resource. The concept of "process function" seems to be problematic. The solution is simple: a process function can also be assigned a capacity (x tasks per hour) and a queue can be placed in front of it. After this we look for resources that are capable of fulfilling this function and that have the required capacity. Just keep the principle of structural indeterminacy in mind.

3.6 Dispersion of Variables and the Required Number of Radar Test Cars

To manage daily business in the flight department, one needs to know the fluctuations that can occur around the averages that have been calculated up to now. This means that the inputs of the black boxes are no longer averages, but are the frequency distributions from Figs. 3.9 and 3.10. The following questions need to be answered.

Questions

D1 Which fluctuations around three airplanes per day in the outputs of both black boxes for "solving deficiencies" should be considered normal?

D2 How many radar test cars should be bought at €250,000 a piece?

Remark
The questions can be answered with the data supplied. The reader is invited to check for himself about how these questions can be tackled. The method would take too much time (about three hours), and is therefore explained over the next few pages.

The model is now complete. The last change was a further specification of the black box "solving deficiencies". In the model, we have so far always performed calculations using the average values of the variables. In order to answer questions D1 and D2, the use of averages is not enough. We should again zoom into one layer; this time not into the black box but *into time*. In other words, we zoom into the behaviour. Instead of working with average throughput times, we now have to work with the real frequency distribution of throughput times of Fig. 3.9. Calculating with this distribution is not as straightforward as with average values. The only solution is to perform a simulation. In the simulation, we follow the course of business in the flight department for several months on paper or on a computer. For a better understanding of this way of calculating, we will first present the way this was actually done in the 1960s, when an appropriate computer tool was not yet available, so the procedure was performed by hand on paper. Section 3.7 will explain the way in which it would have been performed now.

12 radar positions are available. We assume that there are 12 airplanes at these 12 positions. Each time an airplane is completed, we assume that a new airplane enters the position. We now ask what the throughput time of each of these airplanes is, because it can vary between one and eight days. All throughput times should also satisfy the frequency distribution of the dotted line in Fig. 3.9.

How do we accomplish this? We must perform a short sidestep into probability theory.

Assume that six different events can occur. Each event has the same probability of occurring. In this case we can simulate the events by throwing a die. Each face of the die represents an event. If we throw the die enough times, each number will occur the same number of times. The frequency distribution of the events (numbers on top) will then look like Fig. 3.13.

So when there are six possible events we can throw a die. For ten possible events we would need a ten-sided die. Alternatively, tables of random numbers representing the results of throwing a ten-sided die can be used. An example of a part of such a table is shown in Fig. 3.14.

Many manuals contain these tables. The numbers 0 to 9 are equally likely to occur in these tables.

In the case of the flight department, however, the probabilities of events are not equal. Figure 3.9 shows that the probability of an airplane having a throughput time of two days is six times the probability of an airplane having a throughput time of one day. For the radar complaints there are eight different events, corresponding to eight different throughput times (1–8 days).

Therefore we need an eight-sided die, but the die should also be "loaded". When the die is thrown, some numbers should occur on top more often than

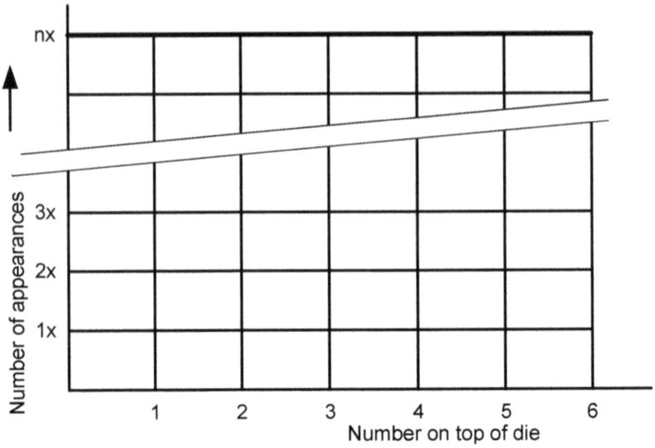

Fig. 3.13 Expected frequency distribution after 6n throws of a die

others. In such a die, the centre of gravity of the die would not be positioned at the heart of the die.

If the frequency distribution were to change a little, the centre of gravity would have to be repositioned. This is far too difficult, so we use a table of random numbers again. If we select two numbers together, we read Fig. 3.14 horizontally (29, 52, 66, etc.), and essentially get a set of possible results from throws of a 100-sided die. Again, the probabilities of each of the numbers 0–99 are equal.

2952	6641	3992	9792	7979	5911	3170	5624
4167	9524	1545	1396	7203	5356	1300	2693
2730	7483	3408	2762	3563	1089	6913	7691
0560	5246	1112	6107	6008	8126	4233	8776
2754	9143	1405	9025	7002	6111	8816	6446
5870	2859	4988	1658	2922	6166	6069	2763
9263	2466	3398	5440	8738	6028	5048	2683
2002	7840	1690	7505	0423	8430	8759	7108
9568	2835	9427	3668	2596	8820	1955	6515
8243	1579	1930	5026	3426	7088	3991	7151
5667	3513	9270	6298	6396	7306	7898	7842
1018	6891	1212	6563	2201	5013	01	
6841	5111	5688	3777	7354	3434		
2041	2207	4889	7346	2865	15		
5565	4764	2617	5281	1879			
4508	1808	3289	399				
2152	6473	5692	0				
6917	4113	7349					

Fig. 3.14 Random numbers

Table 3.2 Classification for the random table of airplanes with radar deficiencies (dotted line of Fig. 3.9)

Throughput time (days)	No. of airplanes	% of airplanes	Random numbers
1	1	4	00–03
2	6	22	04–25
3	5	18	26–43
4	5	19	44–62
5	4	15	63–77
6	3	11	78–88
7	2	7	89–95
8	1	4	96–99
9	0	–	
Total	27	100	

Airplanes with radar deficiencies have eight possible events and the probabilities are not equal. However, we are still able to use the random table by assigning a series of numbers from the 100-sided die to each of the eight events. This assignment should then agree with the frequency distribution of Fig. 3.9. The result of this is shown in Table 3.2.

The first number in the random table is 29 (which means that 29 is the result of the first throw). According to Table 3.2, this corresponds to an airplane with a throughput time of three days. The next throw gives (according to the random table in Fig. 3.14) the number 52. This represents an airplane with a throughput time of four days (Table 3.2).

The simulation of airplanes with radar deficiencies can now be initiated. There are always 12 positions available. The random numbers in Fig. 3.14 correspond to a particular throughput time, as shown in Table 3.2. We can now draw a graph with working days on the horizontal axis and airplanes on the vertical axis; see Fig. 3.15.

To start with, all of the positions are empty. The first airplane has a random number of 29 and thus a throughput time of three days. This is represented in the graph by a horizontal line.

The second airplane has four days, the third five days, the fourth three days, etc. In this way we fill the 12 empty positions on the first day. A thirteenth airplane can now only be added if the repair of one of these 12 airplanes is finished. This finish time can be found in the graph. We look again at the throughput time of this thirteenth airplane and continue. The situation requires some time to stabilise, because we started with an empty hall and the input of the first day was 12 airplanes. The first part of the graph is of little use for the problem. However, stabilisation occurs quite rapidly, and the model can be assumed to be in a "steady state" after the seventeenth airplane; see Fig. 3.15.

There are always 12 airplanes in progress. A vertical line on some working day will therefore always cross 12 horizontal lines representing airplanes in progress.

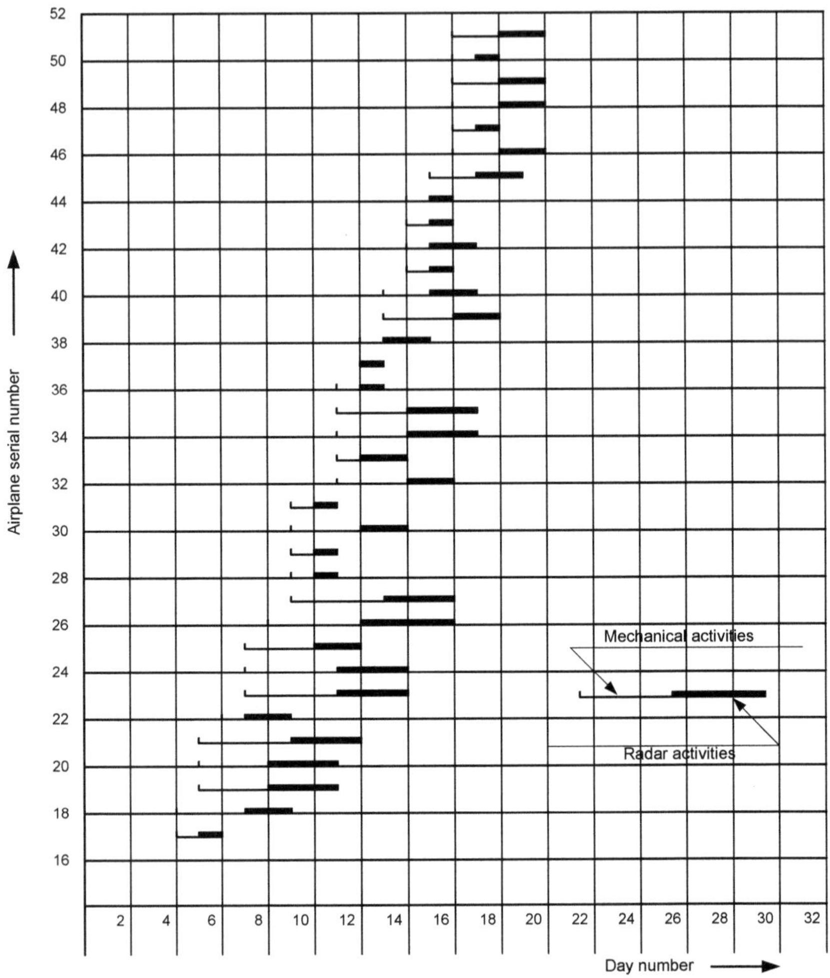

Fig. 3.15 The utilisation of the 12 positions by the airplanes with radar deficiencies

Whenever the repair of one airplane is finished, a new airplane enters the graph. For example, on day 14 the airplanes 23, 24, 30 and 33 are finished and replaced by the airplanes 31, 42, 43 and 44. The throughput time for each of them is again determined by the random table. Table 3.1 also shows which part of the throughput time is spent on radar activities. For example, airplane 18 has a throughput time of five days and according to Table 3.1 the last two days are required for radar. In this way, we can simulate the repair work by hand, say, 300 times.

The output should be on average three airplanes a day. So the 300 times cover 100 working days in reality, which is 20 weeks.

The time required for this simulation is only about three hours for two people. Fig. 3.15 represents the situation inside the black box "solving radar complaints"

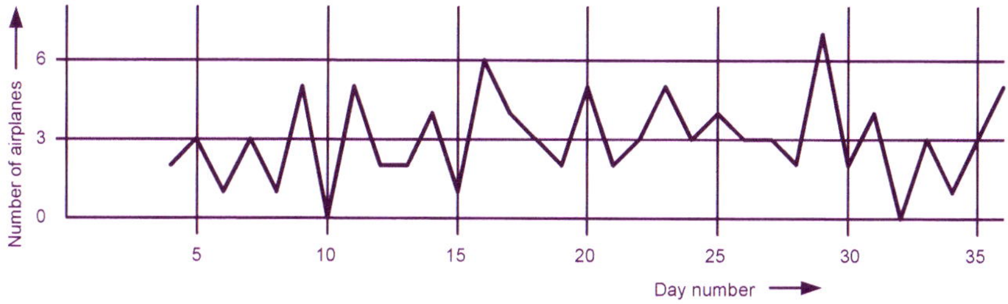

Fig. 3.16 Output per day of airplanes with radar deficiencies (derived from Fig. 3.15; compare for example day 12, etc.)

for each separate day. From this figure, we can also derive the courses of all important variables; first of all the number of airplanes that leaves the black box each day. This is shown in Fig. 3.16. The *average* output should be three airplanes a day.

The graph of Fig. 3.16 shows that seven airplanes are delivered on some days (day 29), and on other days no airplanes are delivered at all (days 10 and 32). The fluctuations around the average of three airplanes a day are quite large, but one should consider it to be normal. This becomes very clear when we draw the input and output in a cumulative way (see Fig. 3.17). The vertical distance between both lines is 12 airplanes, which represents the work in progress. The angle of inclination of the output line equals the output rate. The rate of three airplanes a day is shown by a dotted line.

Figure 3.17 clearly shows that the real output swings around the desired average line of three airplanes a day. If the output drops to zero once every three weeks, then there is no reason to intervene by doing overtime or similar.

In order to determine the number of radar test cars, we count the number of airplanes for which radar activities are being performed on each day in Fig. 3.15. This count is shown in Fig. 3.18. For example, on day 12 in Fig. 3.15, the radar equipment of four airplanes is being repaired. For the other eight airplanes only mechanical deficiencies are being repaired.

Figure 3.18 clearly shows that the 12 positions will never be used simultaneously to solve radar deficiencies. So there is absolutely no need for 12 radar test cars which cost €250,000 a piece. The question is: how many radar test cars are needed? This can be found by finding the frequency of each possible number of occupied test cars (i.e. one occupied test car, two occupied cars, etc.) for the 20 weeks. The results are summarised in Table 3.3.

The situation where the radar deficiencies of ten airplanes are being repaired simultaneously occurs only once in 20 weeks, and can therefore be neglected. The situation where the radar deficiencies of nine airplanes are being repaired occurs six times. These occasions are rather isolated and the peaks are sharp. For a group of two workers it should be possible to deal with these situations by performing overtime. Beyond that, the probability that peaks in bad weather and peaks in radar repair activities would coincide is very small. By using buffers, the need for radar test cars will also decrease a little. It was therefore decided that eight radar test cars would be bought for the positions.

The consequences for the idle time of the employees can also be calculated from these figures. We do not pursue this matter further here.

In this case we have only analysed the functions necessary for the process. We have determined the required capacity for each process function. Therefore, we have only elaborated the "furnishing" of the system. To do this we have only used the theory as described in Chap. 2. A next question could be: "how do we control this process?" The standards and tolerances for the variables are determined already. Only the control loops need to be added to the model.

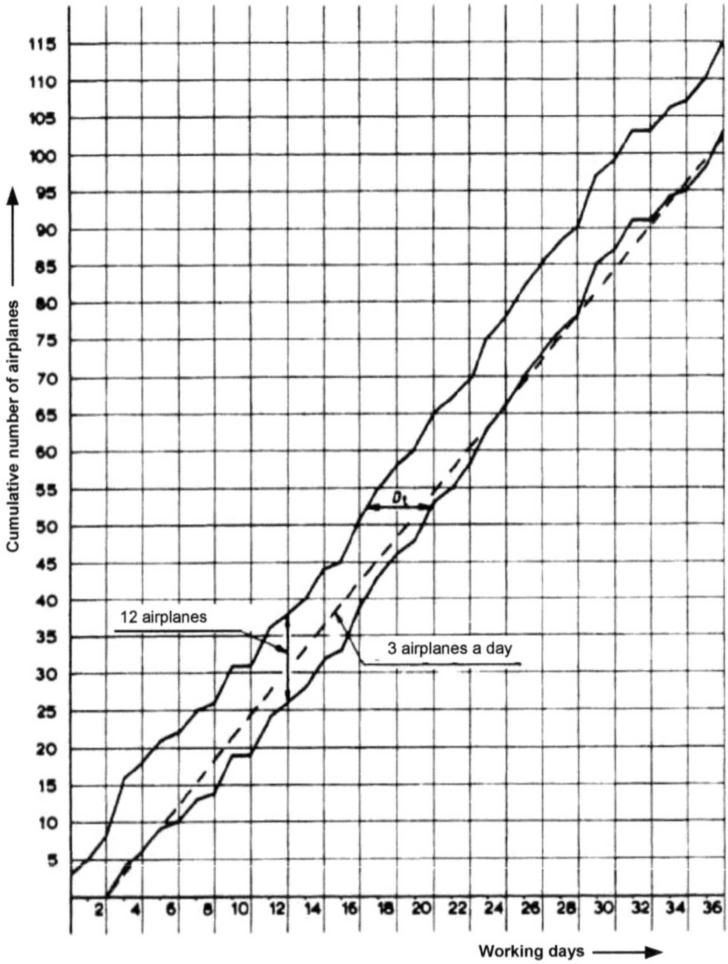

Fig. 3.17 Cumulative graph of the airplanes with radar deficiencies at the 12 designated positions. Dt = throughput time. (Derived from Fig. 3.15; compare for example day 12, etc.)

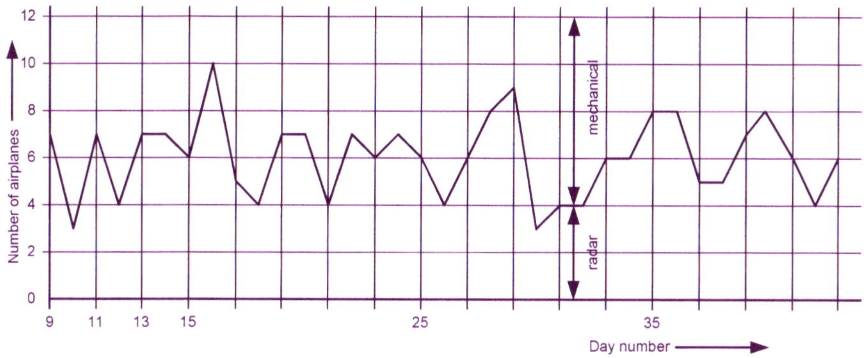

Fig. 3.18 Number of airplanes with radar deficiencies for which the radar is being repaired. (Derived from Fig. 3.15; compare, for example, day 12, etc.)

Table 3.3 Number of days over a period of 20 weeks that a certain number of radar test cars were being used to correct airplane radar deficiencies

Number of radar test cars being used to correct radar deficiencies	Frequency that this number of radar test cars was in use (in days)	Number of airplanes for which only mechanical deficiencies are being repaired
10	1	2
9	6	3
8	13	4
7	23	5
6	15	6
5	15	7
4	18	8
3	8	9
2	1	10

3.7 Results in Practice

In this elaboration of the case, we neglected the consequences of the dispersion of the number of test flights around the average of nine flights until off-test. The troublesome problem that an inspection before a flight is only valid for three calendar days has not been considered at all. In principle the flight department is closed during the weekend and all calculations are based on working days. An airplane that becomes ready on a Friday after solving deficiencies should ultimately fly on a Monday. If Monday is a bad weather day, then the airplane should be inspected again. However, all of these complications can easily be incorporated into the model and solved. The final results of these modelling studies and calculations were presented orally to the management and the customer. There was no time to present a written report. One agreed with the conclusions within a few days, and within two months from the start of the study, tenders were invited for the construction of an extra hall and all equipment was ordered.

The model as developed in Fig. 3.12 was painted on a steel plate. Magnetic disks carried the airplane numbers. One registered the state of affairs hourly from a central point in the flight department. When a magnetic disk left a black box, it was scored on a counting tag at that position. The different graphs like Figs. 3.16 and 3.17 could be updated at night very quickly.

The steel plate model contains the different functions in the process as elements. Within such a process function, the flowing elements—the airplanes—can still be at several physical locations. The model only represents the aspect system of process functions. In order to represent the geographical aspect, another model was required. This model shows where each airplane is exactly positioned on the department's territory. A normal floor plan was used for this purpose. Every plane that entered the flight department was assigned a magnetic disk in the first model and a token in the shape of an airplane in the second model. For both models, one and the same person kept the "score". In order to guarantee that the models agreed with reality, this person had a wired intercom connection with all airplane positions and a wireless connection with all truck drivers. The truck drivers were only allowed to accept from him an order to drive an airplane to another position.

The operations manager of the flight department based his decisions mainly on the state of affairs depicted by the steel plate model until the end of the project. The main cause of this was that this manager had participated intensively in creating the model and had joined the manual simulations; the manual simulations appeared to be particularly probing. One imagines oneself to be part of the simulation. If one decides to do only a computerized simulation, this learning curve disappears completely. The computer tends to distance people from the problem. The procedure described in Chap. 6, where behaviour descriptions can be discussed by everyone involved, may provide a big contribution. The study as presented here and different manual simulations vary only one variable at a time. In order to see what the results would be if all variables were varied at once, a computer simulation is required, although these vary according to known frequency distributions. In the computer simulation, the number of test flights until off-test vary for each airplane; for each flight the number of days taken to solve deficiencies varies; and in addition the number of consecutive bad weather days varies. This cannot be executed manually anymore. Besides this, it is important to pass through the whole course, up to and including the last airplane of the series, in order to get a proper overview of the project. Several computer runs are required for this, because the course will be slightly different each time. An expected delivery zone will originate instead of one delivery line.

The rest of the project took place exactly according to the results of the model. The delivery rate became even faster than the expected zone around the delivery of the two-hundredth airplane, mainly because the number of test flights until off-test decreased, although also because the throughput times decreased a little with experience. When the required production rate was reached, an investigation of whether it was possible to improve the process, particularly the throughput time in the Flight Department as a whole, was then performed. There are two possibilities for achieving this: shortening the throughput times of "solving deficiencies" or trying to decrease the number of test flights until off-test. The effects of these changes were simulated on the computer for different values. One then executes simulations with the model for different *policy options* and looks for the option that has the maximum effect on the results for 350 airplanes. These experiments are *sensitivity analyses* in a way. The experiments showed that decreasing the number of flights until off-test would have the maximum effect and would probably cost the least. It was decided that this option would be selected as the main target. In order to realise it, all deficiencies that occurred more than three times in different airplanes were analysed and work into improving the technical design of the airplane or the test equipment or improving the test procedures was started.

There is one more question that one could ask: why didn't we discuss a more recent situation of a flight department or an analogous situation? The answer is simple. We could only find less complex cases, and in these cases all sorts of refinements in the model would be missing. These refinements are the ones that best illustrate the process of zooming into the function black boxes and into the phase system. The reason that the current situation is simpler is that flight electronics have improved significantly since the F104-G was built. For the F104-G, the MBTF (mean time between failures) of the radar was around three operating hours. This has increased enormously since then. In addition there was a big problem in the coupling (the marriage) between the different electronic systems. When each system operated within its tolerance, there was no guarantee at all that the combination of systems would function within its final tolerance. Often all sorts of extra controls were required which are now not needed. These developments resulted in a far smaller dispersion of several variables than required during the 1960s, and because of this the whole system has become less complex. The F16 for example required only four flights until off-test and a maximum of six flights. As an illustrative example, we still prefer to use the most complex case we know.

References

Little JDC (1961) A proof of the formula $L = \lambda * W$. Operat Res 9:383–387

in 't Veld J, Muller PC (1966) System analysis of a flight department. De Ingenieur, April 22, pp 225–230; April 29, pp 239–243

Chapter 4
Models for the Structuring of Processes

Abstract. In this chapter, a basic conceptual model will be developed for the structuring of processes according to the definitions and systems concepts of Chap. 2. This so-called steady-state model describes the concept of a function (i.e. system) for any repetitive industrial process. The model covers one aspect only that should be selected beforehand but can be applied to all aspects involved. It appears that the system border is not a boundary line, but a border region. Besides this, the function may contain two control regions, the process control and the function control. The one-aspect model will always control the facets quality and quantity, and can be adapted to control other facets too, such as throughput time. The application of the model will be illustrated by a health insurance process.

4.1 Process Types

We distinguish three types of operational processes in industrial systems:

- *Executing processes.* These processes contribute directly to input, to transformations during throughput, and to output.
- *Supporting processes.* They provide both people and resource flows and their maintenance. Not just the maintenance service but also the purchasing of resources and the tools, the recruitment of personnel, training, etc, all belong to this type of process.
- *Controlling processes.* These processes should gear the activities in the executing processes to each other, but they should also gear the supporting processes to the executing processes and gear all internal processes to the environment.

Each process has its own input, throughput and output and is sometimes assigned to separate subsystems.

4.2 Determination of Subsystems

As soon as the main process (the primary process) is established, the functions that should be fulfilled in this process are defined. The activities for one particular process function are often then collected into one subsystem (assuming that this is the best solution).

The concept of a (sub)system implies that there is a border to distinguish it from its environment. However, if we observe a difference in shape, in position, in measurement, etc., between two points in a process, then it does not necessarily mean that these points are logical borders of a subsystem.

Miller and Rice (1967) point out that a true system border implies a *discontinuity*. They hypothesize that this discontinuity on the border shows a change of technology, position, time or a combination of these. This not only holds for the subsystems in the executing process, but also for the subsystems in the supporting and controlling processes. Such a border constitutes, for example, a technologically unavoidable waiting time, such as hardening concrete or waiting for the receipt of an order. The border may also constitute a transfer of a machining operation to a surface treatment. We can also find borders by looking for a narrowing in the flow.

For example, numerous data from different departments are required to write a purchase order. All of these data come together in the purchase order to the supplier. After that a waiting time appears. Over time a payment demand should eventually be made. When received, all kinds of data should be sent to the different departments. We can thus clearly separate the purchase department from goods reception. It is very logical to add the payment demands to goods reception in order to prevent the payment from being demanded when the order has already been received. Beyond that, the purchase order is a clear border point. If we add payment demands to the purchase department, then more data flows are needed through the borders of the subsystem than when it is being done by goods recep-

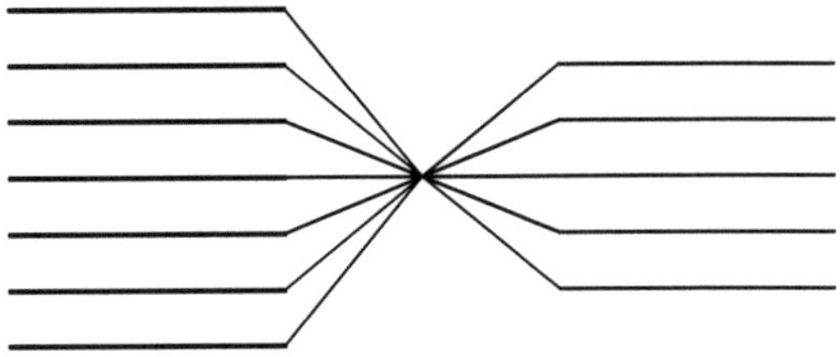

Fig. 4.1a Writing a purchase order

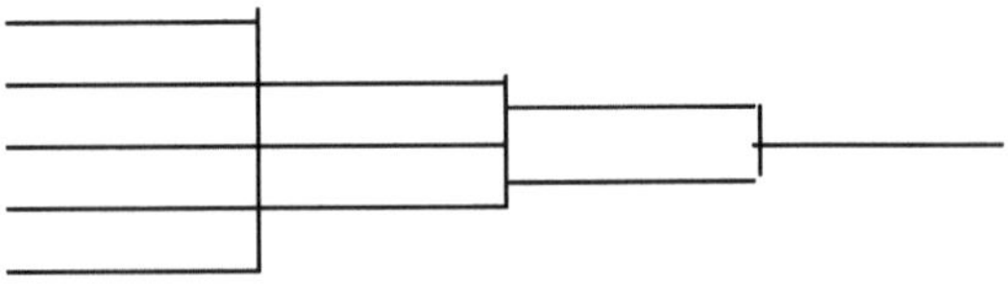

Fig. 4.1b The creation of an assembly

tion. In the latter case, the responsibilities are clearer and the departments are less dependent on each other.

Another logical reason for selecting a system boundary can be a check that needs to be executed somewhere. No matter how much mechanisation and automation have been implemented, there will always be a point where a rest in the process is unavoidable; a rest where checking the output of a process of a (sub)system cannot be mechanized. A system boundary can clearly be defined there.

In other words, it makes sense, depending on the researcher's goal, to select (sub)systems that can constitute more or less autonomous parts. This partition does not have to correspond to the existing partition between departments.

It is important to define the borders clearly, because different people will define them at different positions. This will lead to overlap and differences of opinion or, even worse, to gaps in the flow with numerous entanglements.

A (sub)system keeps its elements together in a meaningful way and maintains its borders via controlling processes. Miller and Rice distinguish:

- Internal controlling processes
- Border-controlling processes.

Miller and Rice call the border-controlling processes "boundary control". These processes appear around and on the system borders and are required to control both the input and output transactions through the borders.

Beyond that, both the input material and the data often need to be reshaped before it can be handled by the system. This boundary control is therefore positioned externally with respect to the executing processes, the real transformations (Fig. 4.2).

In its purest form, the boundary control only allows items that are essential if the goals of the system are to be achieved. In practice this is impossible. There is always a certain degree of border permeability. For example, people are not locked up in the system.

Boundary control is very difficult when there is no discontinuity in the process. The partition may not survive as a (sub)system, because it is too dependent.

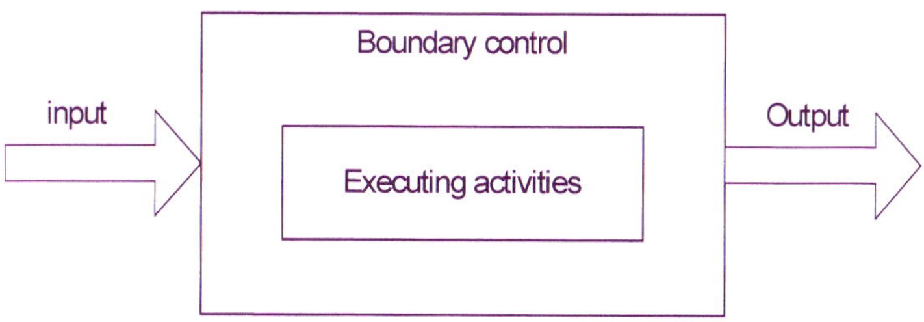

Fig. 4.2 Boundary control

4.3 System Control

Miller and Rice call the internal controlling processes “monitoring”, but this is not correct. In cybernetics, monitoring is only observing; watching the state. It is measuring in order to stay informed, but no interference will take place based on these data. However, within the system measured is also being performed in order to enable control; to process in a controlled manner.

We distinguish four essential aspects that enable a process in a system to be controlled properly:

- There must be an objective; we need to know which output or which state we want to achieve with the system
- The system must be capable of realising this objective
- It must be possible to influence the system’s behaviour in one way or another
- The relationship between the interference and the resulting behaviour must be known.

In order to provide the functionality for all of these aspects, the control functions will now be discussed in detail.

4.3.1 Function Control

The objective here is a direct reflection of the requirements of the system, as depicted in Fig. 2.5. They should be translated into measurable standards for use in the system. This is the goal of the *initiating function.*

The real results of the process are measured in terms of these standards. If the results do not satisfy the standards, the deviation is analysed and the standards may be adapted. The results should again be expressed in terms of the original requirements in order to give higher echelons the ability to judge the performance of the (sub)system. The function that performs these activities is called the *evaluating function.* The combination of the initiating and evaluating functions is called *function control* (Fig. 4.3).

The role of function control in the (sub)system is to translate the rather abstract requirements or goals into concrete standards for the material flow.

The demands that the environment places on the fulfilment of the function can enter the system as clearly defined requirements. For example, the tax inspector, the environment, exactly determines the tax sum that the family system must produce as output for a certain input. The internally defined standards are equal to the externally defined requirements. It is also possible that the system itself must deduce those standards from vaguely formulated requirements from the environment. Then, only information about requirements enters the system, and a function is needed that deduces the standards employed by the internal system from this information.

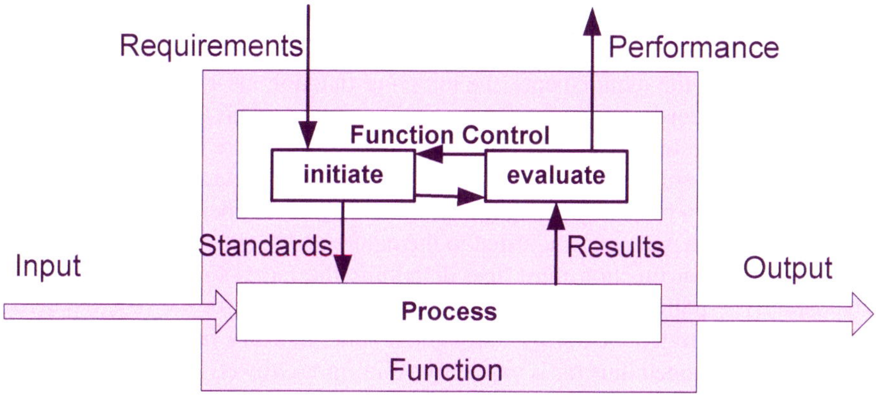

Fig. 4.3 Function control

This usually means a change of time horizon. For example, the requirement for a container terminal is to transfer 500,000 containers a year. This requirement cannot be used as a standard for daily operation. Obviously a cascade of initiating functions (at several levels of management) is needed to translate this into a standard for an operational shift.

We indicate this functionality with the *initiating function*. Actually, this should be an "initiative-taking" function. However, the former nomenclature is so well-established in The Netherlands that it is used anyway. In this thought process, the initiating function also lies within the system in the boundary zone.

The initiating function should preserve the functionality of the system. The requirements are formulated by a higher echelon of control and by research and development. Developments in the technological, organisational or informational field may change the capabilities of the system and therefore other standards may be required.

If the results are regularly insufficient, the requirements may be infeasible, and so maintaining the standards at the same level would violate the second control condition. So function control is also responsible for guarding the "feasibility".

Setting a standard through the initiating function is actually of little use when measurements and evaluations are not made on a regular basis to check whether that standard is (still) correct. *In practice, the standard is only a temporary yardstick.* Changes in the needs of the environment and changes within the system itself may require a change in the standard. Regular control must therefore take place from the external layer to see whether there is still compliance with the given standard and whether the standard, in view of the requirements, is still correctly adjusted. When the standard is continuously not adhered to or when the requirements appear to change, we should check why. This is done via the *evaluating function*. If the standards appear to be incorrect or out-of-date then the evaluating function passes this on to the initiating function. The latter sets new standards and informs the comparison and control functions. In addition, the evaluating function passes on the results, often in a summarised aggregated form, to the evaluating function of the echelon control loop just above.

Starting from the requirements, the initiating function must determine the standards and the decision rules for the executing process. This involves the use of standards for the quality as well as the quantity. Also, there is an exchange between both of these that must be taken into account. Other facets need to be controlled in addition to quality and quantity. These facets and their standards must be deduced from the policy and adjusted to the nature and set-up of the primary process. The initiating function must also determine the facets for which it must set standards. The policy can, for example, provide guidelines for the required effectiveness, productivity, flexibility, throughput time, delivery time, stock position. The model developed thus far is explicitly valid for an *aspectsystem* and the facets of quantity and quality. We can thus work with different (partial) models for various other facets, such as throughput time. As a result, different initiating functions also emerge for the different facets. The nature and set-up of the primary executing process and the policy determine which controlling, initiating and evaluating functions are necessary for which facets. *The existence of an orderly, equipped primary process is crucial in order to limit the complexity of the control.*

When concretely defining standards, we must consider that when the standard is set too low, a slackening usually occurs. If however the standards are set too high, than overstretching, disappointment and finally resignation results. We must never ignore this psychological side of the level of the standard to be set. If the standards are determined by all parties involved in the control functions and in the executing process, then they will more easily believe in them and more readily apply themselves towards realising the standard that they have helped set themselves compared to when that standard is merely imposed by an external body.

Also, when all of the employees collectively determine the standards, then the employees fulfil the initiating function at that moment. They are both workers in and controllers of the process.

The evaluating function serves to compare reality with the standards that have been set. When deviations between reality and standard occur too often, the evaluating function must trace and analyse the cause of those differences. If the standard is indeed set incorrectly then this is passed on to the initiating function, which then reviews the standard. In general, the evaluating function does not respond to a one-off, isolated deviation from the standard.

In addition, the evaluating function must continuously check whether the executing process is still attuned to the requirements of the environment. It thus measures the (internal) possibilities and the (external) desirability of the quantitative and qualitative standards, respectively. In this way, it also establishes whether there is an increase or a decrease in the quantitative needs of the environment, as stated at the beginning of this consideration. If the quantitative needs are reduced to zero—that is, the environment no longer needs the current output from the executing process—then the standards also become zero and the complete process comes to a standstill. The system has lost its right to exist in order to satisfy this requirement. This can happen even when the people and the resources in the system can still fulfil their functions. If we still want the system to continue, we must discover other demands—other goals that can be fulfilled with the same (or adjusted) people and resources. New information about requirements and/or new standards enters the system from functions that are concerned with that development, or the system may even be newly equipped. The initiating function can then also arrive at new standards as a result of the information from the development functions.

4.3.2 Process Control

Function control does not react to disturbances. It assumes an ideal (or best estimate) situation, namely that no exceptional disturbances will take place after the standards are defined. In practice, disturbances will always occur at random or the pattern of disturbances may change (by decay, by external circumstances, etc.), and the system must react to that. The best estimate situation will generally not occur, and in order to deal with this situation, *process control* is introduced (see Fig. 4.4).

Process control consists of two forms of control. The first is *feed forward*; see Fig. 4.5.

Here, the disturbance is determined, after which we compensate for the influence of the disturbance. The disturbance can occur in the input or during the throughput. The disturbance can occur upstream or downstream with respect to the intervention. This should be symbolically presented as drawn in Fig. 4.5. The key point is that the disturbance is measured and that from this measurement the compensatory intervention can be determined. *Cause determines intervention.*

The second form of control is *feedback*; see Fig. 4.6.

The value or state of the output is measured in this control loop. This real situation is compared with the standard situation, in a comparison organ. The real situation

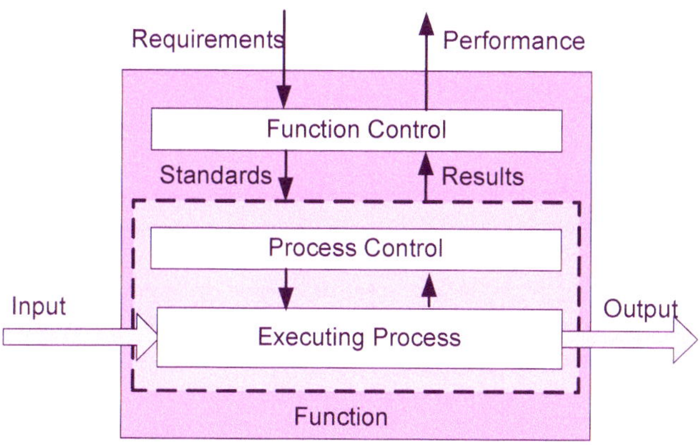

Fig. 4.4 Process control

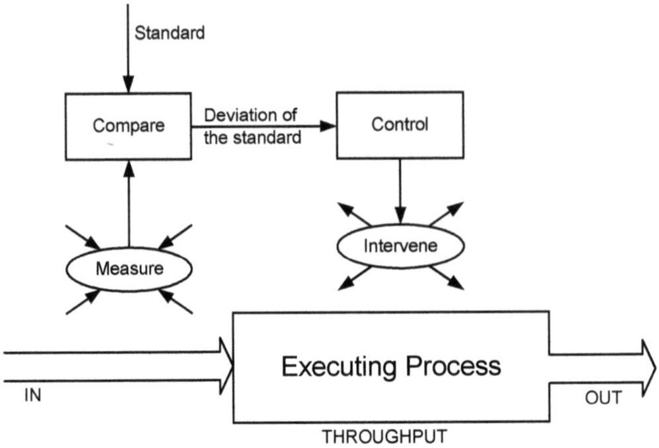

Fig. 4.5 Feed forward, symbolic representation

is the state as it actually is, not as one *thinks* it is. The standard situation is the state as it must be. In the case of a detected deviation between reality and the standard, this information is passed on to a control function. This control function determines the intervention such that we may assume that the output value or the state will adhere to the standard afterwards. This is feedback (reacting). In this way, we also react to unknown and immeasurable disturbances that occur during the throughput. We actually measure the consequence of the disturbance and not the disturbance itself. *Result determines intervention.*

Note the different representations of measurement functions in feed forward and feedback. Measurement for feed forward can take place anywhere in the input or throughput, even in external conditions (like the weather). Measurement for feedback can only take place in the output. This is why the latter measurement is

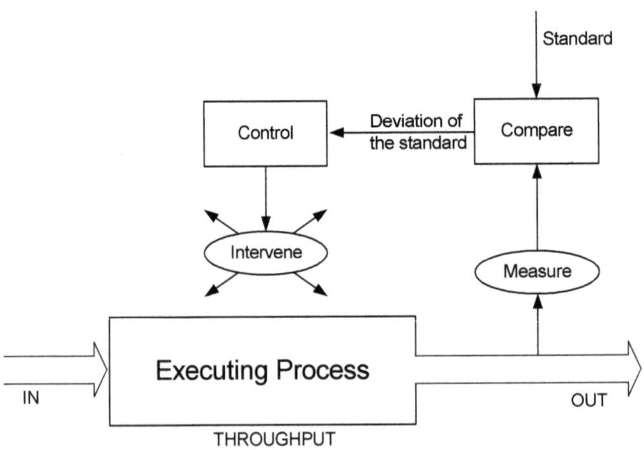

Fig. 4.6 Feedback, symbolic representation

represented by an ellipse with a single input arrow. The intervention function can intervene anywhere in the system, so this is represented by an ellipse with output arrows in all directions.

We generally control at least two facets, namely quantity and quality, and sometimes also throughput time. While it is executing, a system cannot deal with more than is possible with the given persons and resources, or handle it in another manner. Also, the controlling processes themselves have inherent limitations. A system is then also only controlled within its control possibilities.

The measurements, with respect to the feedback, *must* take place at the end of the executing process; otherwise a feedback loop is not drawn over the *complete* process. That is why multidirectional arrows are not drawn to the measuring function, unlike in the case of the feed forward loop; just one measuring *point* is drawn in the output stream. This must then also mean the *state* measurement of the executing process, because the possibility of drawing everything in one model is quite limited. The feedback loop can also carry out interventions in the process or in the resources.

Also, the executing process itself must of course possess standards to which the manufactured product must conform. In addition, there must be standards for the number to be produced per unit of time, the throughput time, etc., and for the required state of people and resources.

A system is *stable* based on certain values of its variables when these variables are inclined to remain within predetermined boundaries. A system can be stable in certain aspects but instable in other aspects. It is nigh on impossible to continuously realise a certain, absolutely, exactly set standard value with feed forward or feedback. Therefore, the set standard will always be a *range* with a lower and upper limit. That is to say, there is a tolerance area. If the interference or the output stays within this tolerance then no deviation is signalled and thus no intervention takes place. In quality management we use random checks and/or statistical methods to determine whether the process remains within the tolerance limits and or whether a shift is beginning to appear. If so, we can take timely corrective measures in the process so that we do not need to wait on actual rejection before we intervene in the process.

We must be careful with the intervention in a system with feedback. It does, after all, take some time (the duration of the transformation process) before the impact of the intervention is measured. The risk of over-compensation and swing-up exists. This phenomenon is firstly dependent on the time required for the total transformation and control process, and secondly on the change in the cause of the deviation over the course of time. We must be aware that this is a dynamic phenomenon, although this is difficult to convey in the static drawings of this book. This can only be discovered through behavioural modelling. In the following account, the reader will be required to imagine certain phenomena flowing through the static, sketched scheme.

The feedback system depicted in Fig. 4.6 is a company. The company's standard output is 100 kg of product per day. The standard is, in this particular case, related to a desired flow rate of the output and not to a state of the system. The

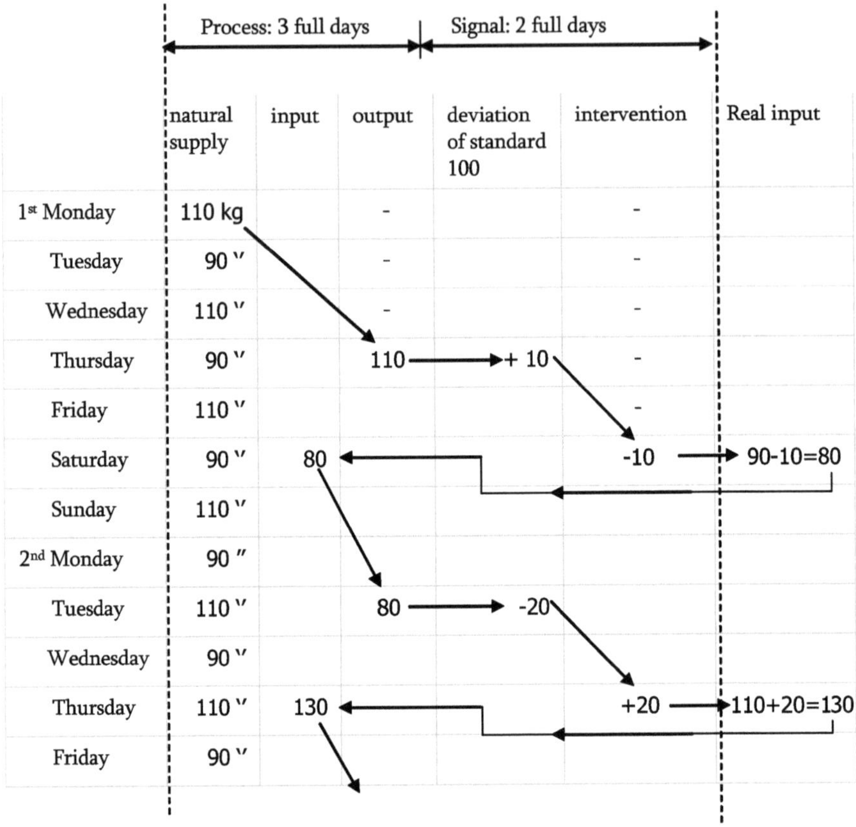

Fig. 4.7 An example of instability by positive feedback

throughput time of the transformation process is three full days. The company operates over weekends too. In order to realise the production output, 100 kg of raw material must be fed in. The desired, constant flow rate of the output (standard) is thus only dependent here on a regular input flow rate. Today, Monday morning, 110 kg are actually fed in due to coincidental circumstances in the supply (see Fig. 4.7).

The consequence of this is that three full days later—Thursday morning—110 kg come out. This is measured by the measuring function (Fig. 4.6) and the comparison function concludes that there is a deviation of +10 kg from the standard of 100 kg. This information is sent to the control function. It takes two full days for this message to arrive at the control function. Based on this message, "10 kg too much is coming out", the control function decides to intervene by closing the input tap by "10 kg". This therefore occurs on Saturday morning. If the input at that moment is still too much, namely 110 kg (the input is *not* measured by feedback) then the action taken is correct, and three days later the intended 100 kg actually comes out.

Now suppose that the input fluctuates regularly, for example on Monday it is 110, Tuesday 90, Wednesday 110, Thursday 90, Friday 110, Saturday 90, Sunday 110, on the second Monday it is 90, the second Tuesday 110, and so forth. The moment that the control function, on that Saturday morning, decides to close the tap by 10 kg, the input is actually 90 kg. The control function does not know this and still closes the tap by 10 kg. The input thus becomes 80 kg. The result of this is that, once again, three days later, on the second Tuesday morning, only 80 kg is emitted from the process. The intervention has amplified the deviation instead of decreasing it. Due to the flow time required for process and control information, the incorrect response is made with respect to the actual situation at the input. That information on the second Tuesday of "20 kg too little" again requires two days to arrive at the control function. This then decides, on the second Thursday morning, to open the tap by 20 kg because too little is coming out. The input is, on that Thursday, actually 110 kg. By opening the tap the input becomes 130 kg. As a result of the flow times and the regularity of the input fluctuation, the tconrol increases the deviation instead of returning it to the standard. In this case, the feedback amplifies the deviation. We call this *positive feedback.* This is summarised in tabular form in Fig. 4.7.

When the deviation signal received by the control function in a system under feedback has the same sign as the deviation occurring at that moment in the input to the intervention function, than one refers to positive feedback, and the deviation can become greater and greater until the system explodes. If these two values have opposite signs then the system returns to the standard. We call this *negative feedback.*

We should not confuse "negative feedback" with "feedback", as is often the case. In a system under feedback, both positive and negative feedback can occur, depending on the parameters.

In addition, it is worth pointing out that in the previous example, this swing-up effect disappears when one reduces the flow time of the signal from the comparison to the control function to just one day. This is the normal reaction in a company when such an instability appears. One deems that the information must be more readily available. In this case, using this combination of flow time and input fluctuation, one can also genuinely stabilise the system, realising negative feedback, by extending the signal flow time to three days. This approach uses not faster but slower and more precise information flow. This system is unstable for this input fluctuation if the sum of the process throughput time and the signal flow time is uneven, but stable when the sum is even.

In the example provided in this section, the cause of the instability was the delay in the signal time. A second cause of instability could be an intervention that is too strong.

Also note that positive or negative feed forward can occur in feed forward due to intervention.

Of course, the measuring point must be able to measure the necessary quantities; however, the accuracy and sensitivity of the measuring point, the measuring function, also play a role. Usually, the less accurate the measurement, the worse the results of the control. An increase in the accuracy above a certain level does not, however, lead to an improvement in the results. That level depends on the properties of the other elements in the whole system. Furthermore, disturbances can take place in both the process and in the signal path. In communication theory we talk about "noise". Noise is an undesirable signal that is generated internally in the system. Both the ratio of the signal strength to the strength of this noise and the location of origin of the noise with respect to the measuring point influence the correct functioning of the process control.

It is theoretically possible to show that noise that originates in the measuring point is more damaging than noise that originates in the control function. Inaccuracies in the measurement of the information are comparable with noise in the measuring point. The accurate capture of the information is therefore more important to process control than the existence of inaccuracies (noise) in the control function.

In general, a separate control loop is necessary for each facet of the executing process that we want to control; the same goes for the functions to be fulfilled in the boundary zone.

The control loops discussed here are always applicable when we want to control a particular facet in the process of an (aspect)system based on standards. These control loops are valid for all facets in the process that we want to control. They control facets in a process in an (aspect)system. These loops are valid for the control of facets such as quantity, quality and throughput time in aspectsystems such as the technological production process, but also in clerical processes, for money flows, space management, personnel training processes, maintenance, improvement in the climate at work, etc.

The combination of function control and process control ensures that the process can repetitively perform in the correct way. This depends, however, on the four control conditions:

- The need that must be fulfilled, the goal of the system, is known
- An equipped transformation function is available that is in a position to realise the target state
- There are ways to influence the system's behaviour
- The relationship between the intervention and the resultant system's behaviour is known.

Function control and process control ensure that this executing process continues to fulfil this need in a controlled manner. The system must therefore have a steady state; that is, the behaviour must be repetitive in time and it must consistently remain focused on the goal, based on a dynamic balance under changing circumstances, equifinality; see Sect. 2.6. From now on, when we speak of the executing process, then we are referring to the process for which we wish to control a certain facet. This process could be product manufacture, the delivery of services, a clerical process, the training of senior management, or the improving the motivation of employees, etc.

Industrial systems and subsystems in industrial systems have goals. These goals are mostly to fulfil certain functions in the environment, the system just above this one. The requirements of this environment must then be translated into useable standards within the (sub)system.

The outputs of the executing processes that actually realise the goals must comply with the standards. Disturbances can occur in the input as well as during the transformation itself and in the output. The output or the state must ultimately comply as closely as possible with the standards that have been set. To that end we must be able to *control* the complete process. The goals are closely related to the aspectsystems to be distinguished. We can usually identify various *facets* (such as quantity and quality) that must be controlled in such a process within an *aspectsystem.* We attempt to set standards for each of these facets, and accordingly enable the process to take place in a controlled manner.

In feedback, see Fig. 4.6, the value of the output or of the state determines the eventual intervention. The throughput time of the process and the signal flow times are important influences on possible swing-up effects. To enable the executing process to occur in a controlled fashion, we measure the output and compare that value with a set standard. In the case of a deviation, an intervention that is determined by the control function follows. To do this, the control function must have a memory and a behaviour model for the system. This feedback control loop is located outside the executing process.

4.3.3 Boundary Zone on the Input Side

The input is often provided in a form that the executing process cannot handle. Therefore, the input must first be made suitable for handling. We achieve this, in its most general form, with *encoding*. For example, it may be that the executing process can only work with the English language. However, the information provided, such as a specification of the requirements, is actually in Spanish. Therefore, in the encoding function, the input must first be translated into English. It may also be the case that the input is composed of great lumps of material but the executing process can only handle fine-grained material. The input must first be ground down in the encoding process.

Each aspect has a quantitative facet and a qualitative facet in its intended result (the standard), and these can be interpreted as the quality and quantity of each single element entering the system or of the repetition of occurrences for the input flow. Both must therefore be combined in one model. The first one is handled in the boundary zone on the input side, and the last is interpreted in process control.

After the encoding, the input is tested in a *filter* against a quality standard that is based on the requirements and possibilities of the transformation function. This test can only take place after the encoding function because we cannot process the uncoded input and so we cannot judge it. When the quality is unacceptable, it must first be brought up to standard. When this is not possible, the input is rejected and returned to the environment. When the quality is high enough, the input is accepted. When the quality is higher than required, the system may choose between accepting it, eventually at a higher cost, or refusing it.

In this filter, testing is performed against a standard and the input is accepted when it lies between certain *tolerances*. The filter chooses between "yes" and "no".

When the quality of the input is accepted, *quantity control* begins. Quantity control can only take place after quality control because unusable inputs would otherwise be counted as being suitable for transformation. Only approved inputs reach the measuring point. The measurements are compared with a standard. If the supply is too large, then the surplus is stored in a buffer so that exactly the right amount enters the transformation. All sorts of intermediate possibilities between yes and no exist, in contrast to the filter. Depending on the standard, the excess created in a period of high supply is stored such that *stock* is generated for leaner times. This buffer function must be located behind the quality filter because it is pointless to set unusable input aside as stock. We should continue to regularly control the quality of the input stored in the buffer. For example, it may decay and thus eventually become unusable in the transformation. This buffer is used to insure against future risks. Some may even use stock to insure against the risk of war or future strikes: this is termed "strategic stock". Sometimes stock is used to create gains on the expected increase in the price of the input: this is called "speculative stock". The existence of stock is always accompanied by unavoidable waiting times for the material. Boundaries are also set for that stock by the dimensions and strength (financial strength among others) of the system. Also, with stocks it is important to judge whether the investment renders enough or covers enough risk to justify the expenditure. When the buffer is full, then the pressure in the supply pipes builds up again until the strength boundaries are exceeded and a crack appears. To avoid this, a *safety* function on the input side is also required that opens when the buffer is full and when the pressure exceeds a certain set value, a standard. This safety function can work automatically but may also be manually operated; for example, when one decides to outsource the excess of work. All of this can be drawn in model form (Fig. 4.8).

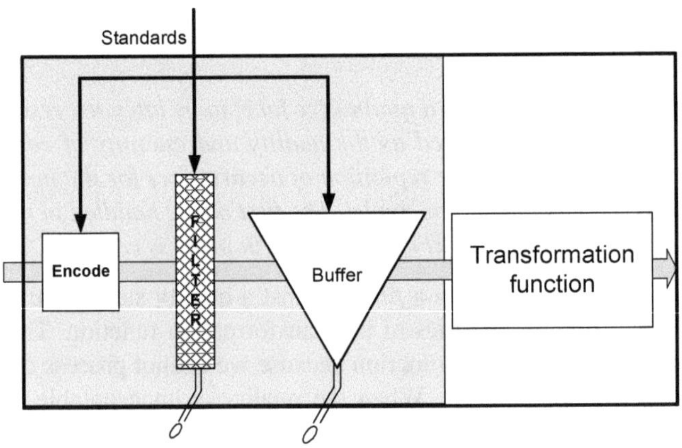

Fig. 4.8 The boundary zone on the input side

All of this occurs before the actual transformation begins. It is a part of the boundary-regulating system that brings the form, quality and quantity of the input to within the values required for the transformation. It is thus clear that for a system we cannot speak of a boundary line with the environment, only a *boundary zone* where the functions described above must be fulfilled.

Thus far, the boundary control has been on the input side.

4.3.4 Boundary Zone on the Output Side

It is the intention that the "controlled" system produces an output or state that complies with certain qualitative and quantitative criteria. This control loop controls both facets. Before we quantitatively measure the output, we should first control the quality of the output in a filter. When the output is qualitatively incorrect it can lead to rejection. The product is then either discarded as scrap in the environment or corrective treatment is carried out to bring the product up to scratch. The product then sometimes passes through all or part of the transformation again. Sometimes the missing bit is added in a separate department. To show this in the model, a symbol for "add the missing" is added to that filter.

When the quantity of the output is too high, the system can often temporarily store the excess in an *output stock* until the environment can absorb that quantity.

Stocks play a tuning role in the total process of input, throughput and output. The principle function of stock is the separation of the input from the throughput and of the throughput from the output. It is thus clearly a *buffering function.* In general, we can accommodate short-term fluctuations in the demand more cheaply in reserve stocks than in reserve production capacity. Stock is a temporary element whereas reserve capacity is actually a permanent element in the system.

In the literature, one only refers to stocks on the output side. On the input side one studies the waiting times. Stocks and waiting times are actually inextricably linked to each other and must be discussed together. We refer to the literature for more on this problem.

On the output side, the chance of an excess (too high) pressure occurring is again present, necessitating a *safety function*. When that opens it can cause difficulties in the environment and for the system. This safety function must receive a standard, just like that on the input side. Further, the possibility exists that the result of the executing process is not immediately suitable for acceptance into the systems of the environment. In that case, an adjustment may still be necessary. We indicate this with *decoding*. These last three functions again lie in the boundary zone. The output boundary zone is drawn in Fig. 4.9. When there is a chance of a disturbance in the decoding it seems advisable to place the buffer and the safety function after the decoding function.

When filters or safety functions discharge something into the environment it will be necessary, in many cases, to report this to the environment. The environment can then take appropriate measures.

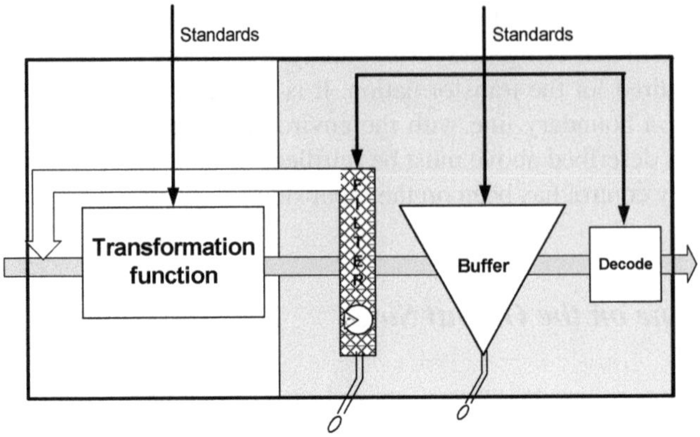

Fig. 4.9 The boundary zone on the output side

We have now further unravelled the content of the boundary zone. The control processes have been positioned with respect to the executing process. The control loops of Figs. 4.5 and 4.6 are symbolic simplifications.

Remark
When the transformation process is lengthy, such as that used by a company, then controlling by feedback alone is usually not enough. In companies and organisations, a combination of feed forward and feedback is almost always applied.

When the transformation process is lengthy and is made up of many different activities, we can split the total process into a series of subsystems. Each subsystem can have its own control loops, as shown in Figs. 4.5 and 4.6. We could place another control loop over a combination of several subsystems. This large control loop is hierarchically placed above the control loops in each subsystem. Mesarović refers to *echelons* here, and a hierarchy of these echelons; see Fig. 4.10. An advantage of such an approach to control (i.e. via a hierarchy of echelons) is that the higher control loop needs to intervene less often and can therefore be simpler because the lower loops quickly eliminate some of the disturbances.

When we talk about feed forward, feedback or boundary control over input and output, then it is extremely important to first sharply define what we consider the

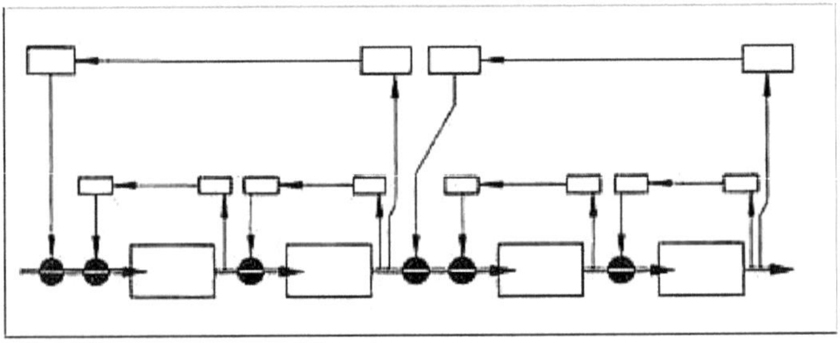

Fig. 4.10 Echelons of control loops

process to be. In Fig. 4.10, the complete process is more or less composed of four part-processes, four black boxes. Now, when we discover a shortcoming in the intermediate products at the end of the second black box, we can rectify this by positioning an "add the missing" at that point, making this part-process "add the missing" too. However, if we look at the complete process, it is "feed forward". The end-product is correct. A disturbance has occurred in the process that has been detected and immediately rectified by an "add the missing" included as a compensatory disturbance. Likewise, feedback applied in a part-process is seen from the viewpoint of the complete process as "feed forward". It is thus of the utmost importance to first sharply determine what we will consider as "the" process.

Of the three processes—executing, supporting and controlling—the supporting process is still missing from the models.

4.4 Supporting Processes

The executing process transforms the input directly into the required output. To facilitate this various supporting processes are usually needed. These could be processes to transport the input, to recruit personnel, to procure resources and maintain them, to improve the people and the resources, to transport support material and energy, etc. These supporting processes could be situated completely within the system. It is also possible that they lie outside of the system's boundary and that only the output of the supporting process enters the system. In this latter case, all sorts of coding and boundary controlling problems could emerge at the system's boundary. To avoid these, wherever possible, it is recommended that the supporting processes should be positioned within the executing process system and they should not be performed by separate systems. This also increases the autonomy of the system.

The control loops for the executing process take daily measurements and take action to compensate for eventual disturbances in that executing process. It was assumed here that the people and resources are a given fact. In practice, the control function usually has the ability, within certain limits, to adjust the people and resources somewhat to the conditions. The control function thus the ability to act correctively by adjusting the inputs, to compensate for the disturbances and/or to adjust the formation and properties of the people and resources used to some degree. (The latter is not indicated in the models). Function control works over a much longer period and only controls the standards. Therefore, we should never interfere in the daily routines of its processes. In the practical application of this model, confusion often arises over these two measurements in the process stream on the output side. This can be avoided by watching out for the reason that the measurements are taken. The purpose of the first measurement is to obtain information that enables the controlled performance of the executing process via feedback. The purpose of the second measurement is to check that the set standards are still correct and improve them where necessary.

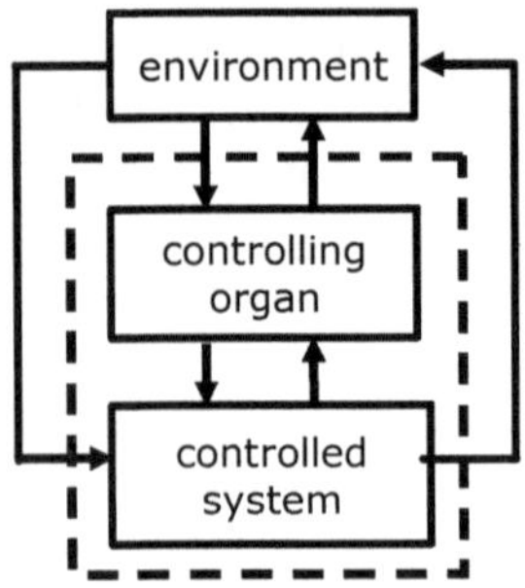

Fig. 4.11 De Leeuw's control paradigm

Many people are aware of the control loops that control the executing process but hardly acknowledge that the standards must, in turn, also be controlled via a standards control loop.

Also, in this function control, the functions involved must possess a memory and a behaviour model in order to select the correct standards. However, a memory is only helpful as long as the elements in the system continue to behave as they have always done in the past. Due to the existence of an evaluating function, the system is capable of learning from experiences and, in doing so, of eventually arriving at a new behaviour model. On the other hand, in practice it is often observed that the original set-up was sound but the elements start behaving differently due to inexperience or ignorance. One no longer knows why one must work in a certain manner or why certain decisions were taken at the time. As a result, one unconsciously deviates from these decisions, causing trouble. One then often reinvents the wheel from scratch. In most companies it is primarily this part of the memory that is neglected and badly organised.

In practice, many managers are inclined to find a (short-term) solution to a problem as quickly as possible. They should really organise *processes* to solve problems and eventually prevent them. In short, they should organise the evaluating function (Sirkin and Stalk 1990).

In this chapter, model development is focused primarily on the functions that must be fulfilled within a system. The control paradigm of De Leeuw (2000) can offer much support, particularly when studying the relationships between a system and its subsystems. This paradigm is based on a system to be controlled, the environment and a controlling organ (see Fig. 4.11). The control possibilities and characteristics are analysed by him. His controlling organ is equivalent to function control. His controlled system is equivalent to process control and executing processes.

Checkland (1998) points out that a system that serves another system cannot be defined and cannot be put into model form before a definition and a model of the system to be served are available.

4.5 The Steady-state Model: Combining the Models into One Model

Figures 4.4–4.6, 4.8 and 4.9 can now be combined into one model; see Fig. 4.12. In addition, the supporting processes must be incorporated. Therefore, Fig. 4.12 provides a function model for controlling various facets of an executing process. It is a function model for one aspectsystem in a steady state with equifinality. Using this, the interlocked facets of quality and quantity can be controlled. Also, the sacrifices for this one aspect are controllable and manageable via similar control circuits.

Comparing and control functions are present in Figs. 4.5 and 4.6. The comparing function only determines the magnitude of an eventual deviation from the standard and passes on these data. When two control functions both implement different or even contradictory adjustments to the system based on different deviation data, there is a great chance that the system will flip over the edge. This is why only one combined control function is present that decides on one particular intervention from all incoming data.

When using this model, it is essential to define exactly what is entered as a temporary element in the aspectsystem, in the flow of matter or in the order flow or the like; for example, the raw material or the order. This gives two different contents to the model because we are examining two different aspectsystems.

The model is universal in the sense that it is independent of the content of the transformation function for whatever type of input. The model is *empty* in terms of the type of temporary elements that must flow thorough the system to be transformed. The model can be applied to achieve a transformation function with an output of coffee grinders as well transformation functions that must heal the sick or produce services.

Not every function that appears in the steady-state model must also always appear in an actual system. If the input is provided in an immediately useable form, then the coding function can be disposed of. When it is it not possible to store the input (for example alternating current), then an input stock is not possible. Also, when the input and throughput rate in a subsystem are exactly matched to a larger system, no buffer is required. If the output product is "conviviality", the output buffer is discarded because that product cannot be stored: it must be consumed immediately. The same applies to many services, such as a seat on a train or a meal in a restaurant.

When the system just above this one—the one that contains the system under consideration—does not lend autonomy to the latter and directly enters the requirements as precisely defined standards in the system under consideration, no initiating and evaluating functions will exist in the system under consideration. These do exist in the system just above in a more extensive control loop of a higher order. The model is therefore not universal in the sense that all of the contained functions must always appear. Case for case, we must ask ourselves whether it is indeed responsible to omit a certain function in that situation. For example, the extremely long process and feedback times that exist in education mean that feedback is of little use. There, we must concentrate on feed forward, but in particular on the organisation of "add the missing" (many years later). When we are confronted with vague goals and lengthy processes, which is often the case, for example, in welfare care work, then that is not a reason to completely omit both of the measurements on the output side and to completely omit an evaluation of the results. In any case, the direction in which the changes must be carried out is known. In all cases, we can reliably measure and evaluate whether the process shows any effect in the desired direction and continues to show improvement in the short term. Obviously, with continuous processes we can never confine ourselves to a single measurement.

In contrast, there could also be more functions than are required by the model. In Sect. 4.7, for example, the need for a distributing function directly in front of the transformation function will emerge. Also, eventual input and output functions are not indicated.

In practice, the steady-state model will just like other similar qualitative models, only rarely be identical to Fig. 4.12. It actually always comes back to the division between executing, controlling and setting standards for processes. The control of a facet always requires one or more control loops and a standards control loop. Furthermore, the initiating function in the standards control loop cannot do without a suitable evaluating function and vice versa. In addition, each control loop must contain a measurement, a comparing function that receives and tests standards, a control function and an intervention function.

The model was completely developed theoretically by starting from the simplest form and then adding more and more functions to account for drawbacks or expected risks. The thought process in the development of the steady-state model is also more or less incremental. The steady-state model will display currently unsolved drawbacks in certain concrete situations. We should add extra functions to solve these. Of course, with each concrete case of designing an organisation, we could start by developing a model from scratch. In the steady-state model, this is actually already done as is minimally necessary for the vast majority of organisations. However, even greater extension of the model produces a strong limitation on that rather general validity. From this model we should in a concrete case progress further and develop a made-to-measure model for that situation. No detailed model can be developed that is applicable to *all* cases. The *law of the situation* is valid here. We do not need to continuously repeat the brainwork up to the steady-state model.

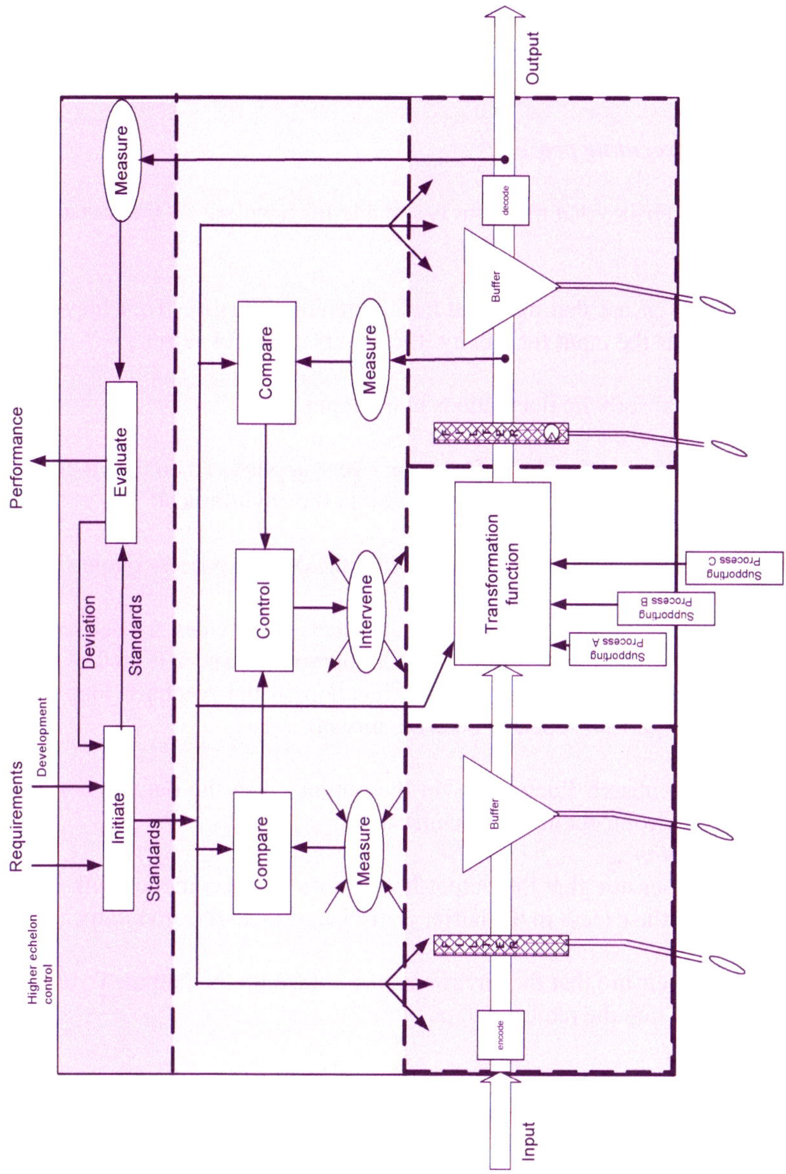

Fig. 4.12 The steady-state model. Function model for one aspectsystem in a steady state with equifinality

Summary

The Functions in the Steady-state Model

Functions in the executing process:
Encoding
The function is: to ensure that the input is suitable for handling by the transformation function.
First filter
The function is: to ensure that the input has the required quality. To achieve this, the function controls the input for quality. It can only accept or reject.
First buffer
The function is: to absorb the fluctuations in the input flow.
First safety function
The function is: to ensure that the buffer is not over-supplied. To do this, it ensures that the excess from a full buffer is discharged into the environment.
Transformation
The function is: to actually convert the input into the desired process output.
Second filter
The function is: to ensure that no imperfect products or services are delivered to the environment. To do this, the quality of the output is inspected (and eventual deviations are passed on to the comparing function in the feedback loop). The function eventually permits repair or adds the missing.
Second buffer
The function is: to absorb fluctuations in the output when the output cannot be delivered to the environment due to circumstances.
Second safety function
The function is: to ensure that the output buffer does not become too full. To do this, it ensures that the excess in the buffer is discharged into the environment.
Decoding
The function is: to ensure that the environment can take up the output. To do this, it brings the output into the required state.

Functions in process control loops:
Measurement for the feed forward
The function is: to measure disturbances in the process within the system's boundaries.
Comparing function of the feed forward
The function is: to compare the values of the disturbances with the given tolerance margin and report the excesses to the control function.
Measurement for the feedback
The function is: to perform post-transformation measurement of deviations with respect to the given standards for all desired facets.

Comparing function of the feedback loop
The function is: to compare the facets of the transformed that need to be controlled with the standards and deviations passed on to the control function.
Control function for the feed forward and the feedback
The function is: to enable the process to take place within the given standards.
To do this, it determines the interventions required to bring this and that back to within the standards.
Intervention function
The function is: to carry out the interventions chosen by the control function.

Functions in the function control loop:
Measurements for the purpose of setting standards
The function is: to measure all sorts of facets of the output and the state of the system over a long period of time.
Evaluating function
The function is: to compare these measurements with the standards and pass on consistent deviations to the initiating function. It must also signal changes in the system's environment that enforce a change in the standards.
Initiating function
The function is: to set standards based on data from a control loop from the echelon just above this level or from the level of development. Or to change standards based on data from the evaluating function.

4.6 Testing a Works Process Planning Department Against Reality

A company makes various metal products according to drawings provided by the customers. These *drawings* enter at the planning department. The head of that department examines the drawings, clarifies them and makes remarks (the coding function). Sometimes he concludes that the company cannot make the product and he returns the drawing to the customer (the quality filter). Should the drawing be accepted, this is registered—the first measurement—and put on a pile (the buffer stock) until one of the planners comes asking for work. If this pile actually becomes so large that the waiting time threatens to become too long, then the boss can still decide to return the drawing to the customer using a lack of capacity as an excuse (the safety function), or he can outsource the preparatory work to a local office. In this last case, he opens the safety function in a controlled manner and allows the input to flow to the following company department via a bypass. That function does not appear in the steady-state model. After a planner has handled the drawing and written out the work assignments (the transformation function), this package goes to the boss who inspects the work (the exit quality filter). Should he discover an error, the whole lot is returned again to the planner concerned with correction, via the re-circulation pipe. Sometimes it appears that the company cannot make the product after all. In this case the drawing still goes back to the customer. When the work order is correct, it is written off as being ready (the second measurement). Thereafter, this package is immediately forwarded to the next department that orders the material and schedules the assignments. There is thus no need for the functions, output buffer, safety function on the output side and decoding. The latter should be unnecessary because this planning department should bring the whole package directly into a useable form for the next department in its transformation function. This is, after all, its assignment. The planning department as a whole fulfils the coding function in the system just above this one, the complete company.

Every day the boss updates a graph of the number of drawings received and the number of finished work orders. Using that graph, he can determine the average throughput time between both measurements. The works manager (environment) has stated that this can never take more than one week. Therefore, initiating and evaluating functions do not appear in the planning system. The boss actually fulfils the functions of comparing and controlling by testing his throughput time graph against that standard of one week and permitting eventual overtime or outsourcing if that "week" threatens to be overrun. It is also possible that the boss finds the week as stated by the works manager to be a completely senseless demand and does not use that as a standard at all. The coding function in the standards control loop then rejects the standard and does not allow it into the system. The boss receives requests from the next department to comply with certain wishes regarding the content of the work orders. He then starts to employ these as standards at the output test in the second quality filter. There is also the possibility at this point that he finds the request to be absurd and does not admit it as a standard in the system.

In this case, the boss himself fulfils all functions outside of the transformation function yet within the system's boundary. It is also possible that, for example, a special employee fulfils the final inspection, the second filter.

This is a short example of an interpretation of an actual system in relation to the steady-state model.

4.7 Nurses Effect

Previous considerations and the steady-state model are valid for every executing system, small or large. For each aspect that must be controlled, functions and control loops, in accordance with the steady-state model, are essential. Up until now there have been no models that show the interrelationships between the different aspectsystems.

In this section we will examine the entire complex system as a black box for our first study. Then we will open the black box a little, but still keep the transformation function closed. We actually make the cover of the black box a little transparent so that we can only see the boundary zone only. We refer to this step as "the *greying* of

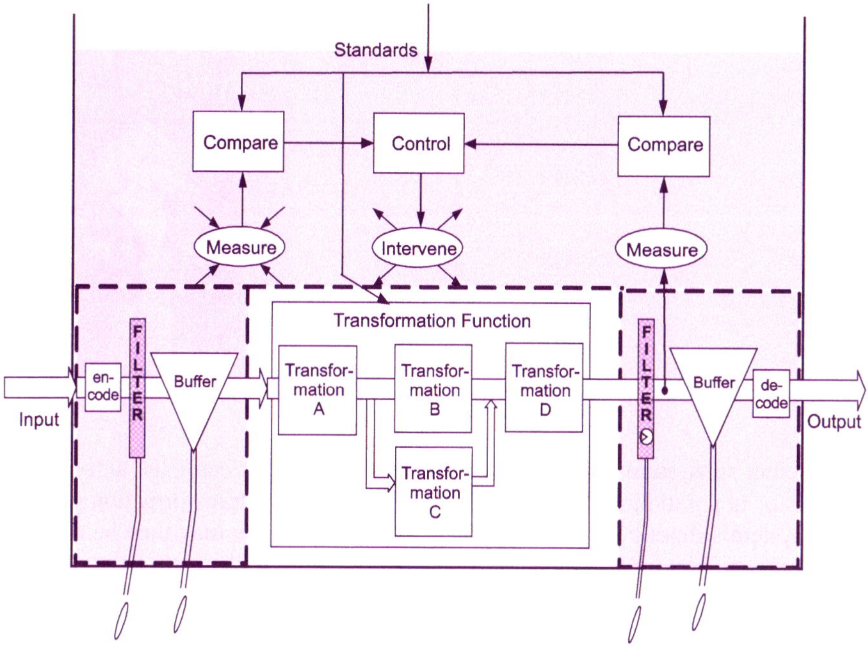

Fig. 4.13 Subsystems, in series and parallel, within the transformation function

the black box". We do not actually zoom into a complete aggregation layer; only a little bit of it. This is the situation in the steady-state model. The black box "transformation" function is only opened and the whole aggregation step completed when more details appear necessary. In general, we then encounter all sorts of subsystems A, B, C, and D in series or parallel in that box, as shown in Fig. 4.13.

Again we can see each of these subsystems individually as a system that we first approach as a black box before eventually using the steady-state model. In turn, processes are also required in the coding function, in the filters, etc., with their own boundary zones and standards. This model, where we first zoomed into one aggregation layer, is therefore valid once again, just as it was for the supporting processes, as indicated by a rectangle in the steady-state model. Again, this demands an "executing" process with its own input and output, its own boundary zones and its own control loops and standards control loop.

That same steady-state model for the aspectsystems is valid in principle for each function in the steady-state model when we zoom into one aggregation layer. The picture shown on the tins of a well-known brand of cocoa, see Fig. 4.14, nicely illustrates this. The nurse carries a tray bearing the same cocoa tin with the same figure on it, etc. We call this *the nurses effect*. At each aggregation layer we encounter new coding functions, control loops, etc. In this way, a hierarchy of systems appears to exist. Each system is composed of subsystems and is itself a part of a suprasystem, the environment. Each subsystem is dependent on its environment, the system, and derives its primary objectives from it.

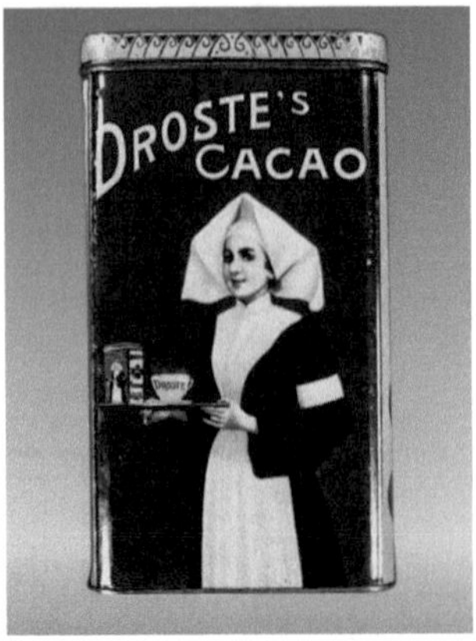

Fig. 4.14 "The nurses effect" (SFA)

Different subsystems at an equivalent aggregation layer can exist side-by-side in series or in parallel, and they each execute a part of the transformation process. In the system's input boundary zone, a *distributing function* may then be required that directs the input into the correct subsystem. In turn, each subsystem has internal control loops. These control loops, which are all at the same aggregation layer, must be interconnected. This is generally done by the control loop in the aggregation layer just above this one in the system. That loop is of a higher order and encompasses a number of these subsystems. Thus, a hierarchy of control loops also exists. We refer to these as echelons. That hierarchy and the span of each loop are connected to the number of variables that must be involved in the decision-making process and with the level of abstraction of the decisions to be made. When the data required to control a certain subsystem is not present within that subsystem, a control loop from a higher echelon that does contain the data must be established. This may also be necessary when the ability to handle the data and the ability to make a decision are not available in the subsystem; for example, through insufficient knowledge and education of the employees present. A third reason for setting up a control loop on a higher echelon could be that the decisions in the subsystem could produce external effects that influence the functioning of other subsystems to the detriment of the system as a whole.

We must not confuse a control loop from a higher echelon that controls the daily output with the control loop for setting standards.

When the subsystems are arranged in series in the transformation process, the output stock from the previous subsystem also forms an input stock for the following subsystem. The boundary zones of the subsystems can overlap to some extent. The control loop of a higher echelon will also devote attention to these intermediate stocks and to transport between the subsystems. Even if we also consider the transport to be an "activity", it appears that, in many organisations, the product actually spends no more than ~8% of its total throughput time in actual processing. The intermediate stocks in the pipeline are usually large and it is important to attempt to reduce these considering the space, costs and risks involved. This problem is often even worse for service and clerical processes.

Hammer (1994) mentions an insurance company where the total throughput time for a policy was 28 days, only 26 minutes of which were actually spent working on the policy. That is 0.19% of the throughput time!

The types of boundary-controlling processes at the borders between the subsystems provide an opportunity to determine whether the division between subsystems has been located correctly. When we place a border at a point at which there is a large discontinuity in the process then the boundary control processes are simple. As a pause is required anyway to code and/or to inspect the preceding work, these boundary-controlling processes are also efficient. When there is only a very small discontinuity then a far more extensive control apparatus is required: firstly to maintain the differences between the two subsystems and secondly to ensure coordination between them. Sometimes an extra border will need to be created artificially because the control methods and/or means available cannot span such a large subsystem. This can be the case when there is a large geographical spread of the elements. This is a border resulting from limiting conditions.

Each control loop can only control a given process within its control possibilities. These possibilities are partly determined by the properties of the control loop itself (for example, capacity and competences), and partly by the maximum capacity and the properties of the people and the resources in the executing process. The technology and the organisational structure are also resources. We should therefore look ahead at the control loops and determine whether new requirements will override the existing control possibilities in the future and which measures we must then take. The period over which we look ahead is separated into:

- Short-term planning (three months to one year)
- Mid-term planning (one to five years).

The short-term planning is relevant primarily within the smallest control loops. In the control loops at higher echelons, the period over which we look ahead becomes increasingly longer. This also means that this usually occurs at a higher level in the company's hierarchy. In many situations, management and so-called staff units also fulfil, in addition to the functions in the boundary zone, the internal control loop of each system in each echelon. Each echelon complies with the steady-state model, but at a higher aggregation layer.

This all still departs from a set of given needs and an existing, equipped transformation function. It explicitly concerns the executing processes as well as their control and scheduling for a determined existing need, corresponding to the steady state and equifinality. What should be done, however, when the evaluating function establishes that there is no longer a need for the output of this existing executing process? None of the functions discussed thus far have, as a goal, the continuation of the system. No single function is focused on change and renewal. The potential for innovation, for breakthroughs, is not present. In Chap. 8, we will devote our attention to this major drawback of the steady-state model developed here. However, you will now apply the steady-state model using the case described in the following paragraph.

4.8 Case: The Health Insurance Company[1]

This case is about a clerical process.

XYZ Insurance is a regional health care insurer. In addition to a variety of other types of insurance, the "implementation of the national health law" is one of the main tasks of this insurer. Within XYZ we only consider the process from the handling of a request for a new health insurance policy. The policy control department does this. The principal objective of this department is the registration of applicants for national health insurance and the maintenance of policy data. Twenty people work in the department, spread over three regional groups. Each group is responsible for maintaining the data of those covered by XYZ national health insurance in their region.

Preliminary Assignment

Draw, on the basis of the following description, a process model with control loops, for as far as they are found. Look through a "steady-state pair of glasses" at the described reality. Indicate clearly where you would draw the system's boundary and what the flowing element is. Start an aggregation layer where the *actual transformation* from a completed application form to a definite policy is one black box. Then, for as far as the text permits, zoom into the black box and draw all of this in one model.

The requests for registration for health care insurance arrive in the post room. The envelope is opened and someone then decides which region group the request must be forwarded to. Then of course, the request is sent internally.

In the particular region group, the date of receipt is first stamped on the request form. Then all details on the request form are inspected for clarity and completeness. Should items be missing, the form is sent back to the applicant with a request to provide the missing information. When the request form is resubmitted, the steps described above are repeated.

Thereafter, based on the details supplied, someone checks whether the requested insurance package is among the packages offered by XYZ and whether the applicant conforms to certain acceptance criteria. If not, the request form is returned to the applicant together with an explanation. When the form is complete and the request appears to be acceptable, the data are given certain code numbers, depending on which general practitioner, dentist and pharmacy have been cited. In addition, a code number is added to indicate which insurance package the applicant wishes to receive.

The policy terms are then determined by a "mutalist". Sometimes this person discovers that a code number is missing. He adds this anyway. Considering the

[1] This case is based on a report of practical work performed by Mr. S. Rep, student at the Institution for Economic Studies at Amsterdam.

fact that determining the policy conditions vary somewhat in terms of time required and the preceding process function of assigning a code number only takes a little time, the forms usually enter a waiting stock. When this waiting stock is empty, the mutalist rings the coder to encourage him to speed up a little.

Subsequently first the mutalist himself and then also the insurance applicant must control whether they are in agreement with the conditions: have any errors been introduced into the data? Are the policy conditions and the cover what the applicant actually requested? And is the premium acceptable? This quotation must be sent to the applicant and is received back from the applicant with eventual approval or desired changes. In the meantime, a copy of this quotation lies in one of the mutalist's in-trays. The post room sends the applicant's response directly to the mutalist. The mailing date and the date of receipt are noted on the copy of the quotation. The final policy with conditions and proof of insurance can then be compiled and sent. Should errors be discovered in the quotation, then these are redressed and an adjusted quotation is re-sent to the customer. Although this is not the intention, it is possible that this cycle is repeated a number of times.

Before dispatch, the various data are always entered into the computer and the total throughput time is recorded. Management has stated that the total process must not take longer than 30 calendar days. All the same, the mutalist checks everything for data accuracy and conditions, etc., before sending the final policy.

Someone also regularly compares the contents of the XYZ policies and conditions with those of other insurance companies. The insurance packages offered by XYZ and the policy conditions are also regularly checked to see if these (still) comply with the requirements of management and the law.

Assignment

Based on what has been discussed, draw the process models at two aggregation layers (see previous text) and the eventual control loops. In a separate overview, describe which functions are fulfilled and by whom. At which points does it seem that further investigation and/or supplementary data are necessary?

Solution

The first model is given in Fig. 4.15.

The post room is placed outside the system's boundary because it does not to be required for an analysis of the process. It only fulfils a distributing function. The customer is also excluded from the system. We consider one regional group as the system, as both of the other groups yield an identical model.

The input is "a request for heath care insurance cover" and the output is "a prepared policy".

Therefore, at the first presented aggregation layer in the assignment, we find only a coding function, a quality filter, the transformation function and eventually the output quality filter. Perhaps you have already drawn in control loops.

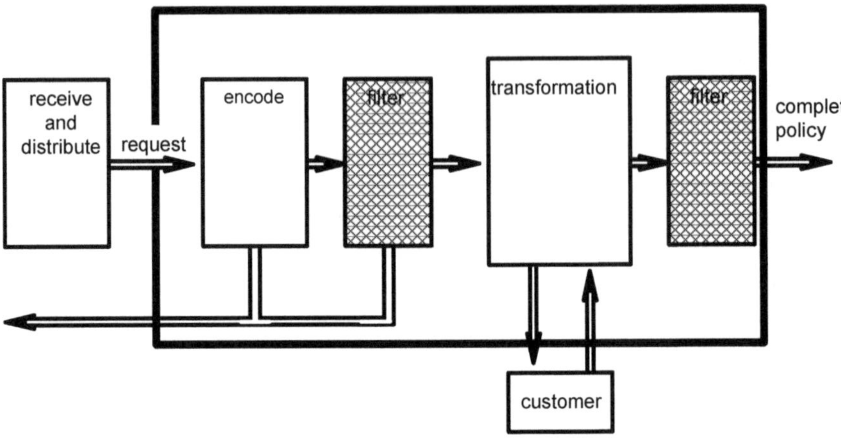

Fig. 4.15 The first model

More detailed data have been given for the transformation function in Fig. 4.15. We can thus zoom in another layer on that black box.

In Fig. 4.16, the eventual model with the following functions is drawn:

1. *Encoding function:* Prepares the request for handling by the transformation function by controlling the presence of all data. If incomplete, the application is returned to the sender with a request to provide the required information.
2. *Filter function*: Controls whether the insurance package requested is offered by XYZ and whether it adheres to the XYZ acceptance criteria. If the application does not meet these criteria then it is returned, with an explanation, to the applicant.
3. *First transformation function*: The various data are given code numbers. (The fact that one refers to code numbers does not mean that this is a coding function! It is a necessary activity. We only draw it as a separate transformation function because a buffer functions in the process after this). It is not clear who should fulfil this function from the text. That's why there is a question mark on the lower, right-hand side in the black box.
4. *Buffer*: Now the form arrives on a pile belonging to mutalist A or mutalist B or mutalist, etc.
5. *Second and actual transformation function*: The mutalist now sets the policy conditions and notes these in a quotation to the customer.

Feed forward: There is talk of feed forward in two cases here. Firstly, when the waiting stock of the mutalist is zero, he acts by making a call. He ascertains a disturbance and attempts to compensate for this by performing an action. Secondly, he adds a missing code number there and then. He thus compensates for a disturbance. We could see this as "add the missing", but that is incorrect. At the layer of the model drawn, which encompasses the *total* process, it is a compensation for a disturbance during the process. It is only "add the missing" when we look at a partial process.

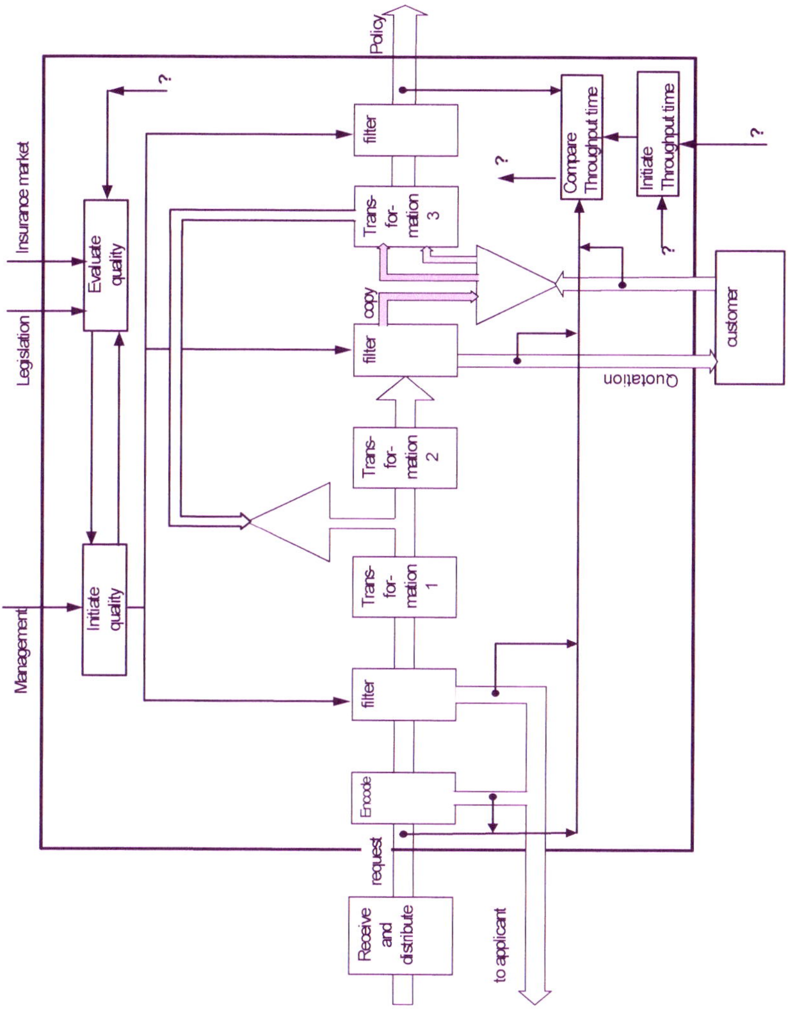

Fig. 4.16 The second model, where we have zoomed in by one aggregation layer

6. *Quotation output filter*: Prior to sending the quote, this is checked again by the mutalist. He retains a dated copy and stores that copy in a buffer (function 7).
7. *Buffer*
8. The customer also fulfils a quality filter function by controlling whether the quote is both correct and to his/her satisfaction. The customer however falls outside the system under consideration.
9. *Third transformation function*: On receipt of the customer's response, the mutalist retrieves the duplicate quote. If adjustments are required, then the previous cycle is carried out again. If the quote is accepted, then the mutalist can compile the final policy.

10. *End filter quality*: The mutalist checks the final policy once more, after which it is dispatched, probably via the post room that is however kept outside the system. The post room can be alternatively regarded as a decoding function.
11. *Quantity measurements*; *in this case, throughput time*: The date on which the particular activity takes place, at different places in the process, is entered into the computer. See the measuring points in Fig. 4.16.
12. *Comparing function for the throughput time:* Various calendar dates are entered into the computer, which will, in turn, do something with them. However, apparently, only the total throughput time is compared with the standard "30 calendar days". There is, however, nothing in the text about what should be done with those data. It is unclear whether there is a control function and whether interventions take place if the standard is exceeded, let alone whether that standard is evaluated.
13. *Initiating function for the throughput time:* This is apparently present, namely as the 30 days that management has set.
14. *Standards setting control loop for the quality:* The quality filters incorporated into the process receive standards for the acceptance criteria and the content of the packages to be offered.
    These standards come from an *initiating function* (15). This determines the standard based on guidelines from management and on data from the evaluating function.
    The evaluating function (16) regularly compares the content of the XYZ packages and conditions with those of other insurers. Furthermore, it responds to amendments in the law. Deviations are passed on to the initiating function. It is, however, unclear who actually fulfils these functions from the text.

Remark
The abovementioned function numbers are reported bottom left in the black boxes of the model. Who fulfils that function, such as Ma (mutalist A), is indicated bottom right. If this is this unknown, then a question mark is present.

The model thus evokes a number of different questions that would justify further research. Who fulfils the functions in the standards control loop for the quality? In the daily control loop for the quality it appears that only the mutalist checks himself, but who checks the mutalist? Also, the daily control of the throughput time and the intermediate monitoring of progress are unclear. Not a single word is mentioned about the total throughput capacity. The mutalist fulfils all sorts of functions, but who does the coding and the addition of the code numbers?

4.9 Some Applications in Practice

The steady-state model has been applied in the analysis of organisational problems in many areas in industry, as well as in banks and in the service sector. Reports on such analyses are usually only published internally and are therefore not generally available. Only a few examples will be briefly discussed in this paragraph. For more detailed information, please consult the references. The following examples have been selected from completely different areas of application.

The steady-state model is developed with an eye to application in organisations. It has been found to be a good tool, when used in combination with thinking in terms of functions, even for the automation of a purely technical process.

Lohmann (1990) applied this when designing the software for an automated dredging process. There were primary functions, and within each of those primary functions, subfunctions could be distinguished. The primary functions set standards for the subfunctions and transmitted these to the subfunctions. Each subfunction performed its activities and checked these with the standard set by the respective primary function. In the meantime, the primary function could actually change the standard, and the control loop in the subfunction should then react to this. When the design team involved met with problems, they were usually traceable to "not logically thinking in terms of diverse layers" and it appeared that, whilst thinking, someone confused a control loop with a standards control loop. The respective programme in question took approximately thirty man-years and worked flawlessly at the customer's final inspection. This model has also begun to find application when designing company or management information systems. It is also applied in accountancy.

Plasschaert (1983) applied the model to the organisation of a dentist's practice. He clearly illustrated the function concept using the process of which the dentist is a part. The control loops in a dentist's practice, used for controlling quantity and quality, are also worked out practically.

The steady-state model was also applied to a doctor's practice and to a group practice.

The model has proven its worth in quality management and in the framework of the introduction of the ISO-9000 standards. Van Dijk and Seegers (1991) made use of the thought process in the development of a computer-steered quality management system. It is a good aid when analysing the organisational structure required and when determining the ISO-9004 procedural descriptions required, etc. In 2000, a new version of the ISO-9000 series was published. This was a great advancement. Now, attention must be paid to the control of the processes and customer satisfaction, and a management audit must be performed regularly. In this way, the new ISO-9000 standard better fits thinking in terms of processes and control loops. Also, in turn, the ISO-14000 standards for environment management better match the methods of the new ISO-9000 series.

Leenders and Langemeyer (1990) used the steady-state model in an integrated approach to the reorganisation of a production process at DAF Trucks, namely the welding of sheet metal parts.

Also, van den Berg (1993) used the steady-state model and the total model (discussed later) in his book entitled *Kwaliteit van levensmiddelen* (*Quality of food*).

An important application of the steady-state model is the software package EXCOM from TLO-Holland Controls. It is a computer-controlled measuring instrument for the quality of labour at one job. A job analysis is made, aided by a long series of questions to the worker. Thereby, all of the functions of the steady-state model, in addition to the working conditions, are broached. Based on sociological and socio-psychological research, it was accepted and incorporated into the computer programme that a high-quality job must contain all functions of the steady-state model. The answers to the questions are summarised in five thermometers: purposeful, autonomy, participation, work content and working conditions. Furthermore, the steady-state model appears on the screen so long as the functions concerned are encountered at that job. Actually, in this way, the job is judged as "qualitatively high" when all functions from the steady-state model, with the appropriate authority, are encountered in that job, there is enough influence on adjoining jobs, and the working conditions are good.

Various large companies use this software package to analyse their existing jobs (positions) and design new structures and new jobs. Both the management, as well as the employees involved, are enthusiastic about the results. Use of the instrument repeatedly leads to interesting discussions and improvements.

Summary

Based on theoretical considerations, a function model for a system was developed (the steady-state model, Fig. 4.11) in which feed forward and feedback are present for the executing process and a standards control loop appears for the standards against which the feed forward and feedback must be checked. This model was elucidated using "a process planning department", and, as an exercise, the case for the insurance company was analysed and placed into modular form. The model is so universally applicable because:

- It is "empty" in terms of what flows in it; they could be teaspoons, patients or services
- It can be applied again at each aggregation layer (the nurse's effect).

Due to this latter effect, it appears that both the systems and the control loops form a hierarchy of echelons.

References

van den Berg MG (1993) Kwaliteit van levensmiddelen (Quality of food). Kluwer, Deventer

Checkland P (1998) Systems thinking, systems practice. Wiley, New York

van Dijk FMT, Seegers JMM (1991) Computer-ondersteunde kwaliteitszorg (Computer supported quality care). Metaalbewerking, April 1991, pp 122–128

Hammer M (1994) Een chef werkt niet, hij plant. De Ingenieur, Dec 1994, pp 6–9

Leenders PWJ, Langemeijer FL (1990) Automatisering lasbewerking: integrale studie noodzakelijk (Automating welding operations: integral studies required). Werktuigbouwkunde (Mechanical Engineering) 5:17–42

de Leeuw ACJ (2000) Bedrijfskundig management (Business management). Van Gorcum, Assan
Lohman TAM (1990) EXCOM (Report S86). Directorate-General of Labour, Den Haag
Miller EJ, Rice AK (1967) Systems of organization. Tavistock, London
Plasschaert AJM (1983) De tandartspraktijk als organisatiesysteem (The dentist practice as organisation system). De Tandartspraktijk (The Dental Practice), Oct 1983, pp 36–39; Nov 1983, pp 10–13
Sirkin H.S., Stalk G. (1990) Fix the process, not the problem. Harvard Bus Rev, July/Aug 1990, pp 26–33

Chapter 5
Conceptual Model for the Analysis of Industrial Systems

Abstract. The steady-state model is defined as the basic model of the Delft Systems Approach. However, it only models one aspect at a time. When starting an analysis of an industrial system, one always should consider at least three aspects and their interrelations: the material flow, the order flow and the resource flow. To do this, in this chapter we introduce the PROPER model, which consists of three steady-state models for these three aspects. PROPER stands for "PROcess-PERformance", because we consider that the systematic preservation of this combination during the analysis distinguishes it from other approaches. In the first part of this chapter, it will be shown that other conceptual models only show control relations; they leave relations to real operational processes poorly defined. Finally the conceptual PROPER model will be applied to the well-known reference model SCOR.

5.1 Introduction

In Chap. 4 we introduced the steady-state model and the control paradigm as conceptual models for a purposive system. Industrial systems belong to this class of systems. In this chapter we will first investigate other conceptual models for these systems, in order to prove the validity of the steady-state model on the one hand and to highlight the additional value of this model on the other. A combination of function and process will be shown to be the unique property of this model. Then we will construct the PROPER model for the analysis of industrial systems, which is based on the steady-state model. Being a conceptual model, we will apply this concept to a widely used reference model for logistic systems: the SCOR model. We will show that the conceptual model is able to represent this reference model, but additionally offers the opportunity to represent other (and more innovative) systems.

5.2 Other Conceptual Models

5.2.1 Formal System Model

Checkland calls a conceptual model based on the soft systems methodology a formal system model (FSM). Macauley (1996) defined a FSM for a system with human activities. This model is also called a HAS (human activity system) model. According to Macauley, the system S is a formal system if and only if it meets the following criteria:

- S must have some mission
- S must have a measure of performance
- S must have a decision-making process
- S must have elements which mutually interact in such a way that the effects and actions are transmitted through the system
- S must be part of a wider system with which it interacts
- S must be bounded from the wider system, based on the area where its decision-making process has the power to enforce an action
- S must have resources at the disposal of its decision-making process
- S must have either long-term stability or the ability to recover in the event of a disturbance
- Elements of S must be systems with all of the properties of S (except the emergent property of S)

Such a formal system is presented in Fig. 5.1.

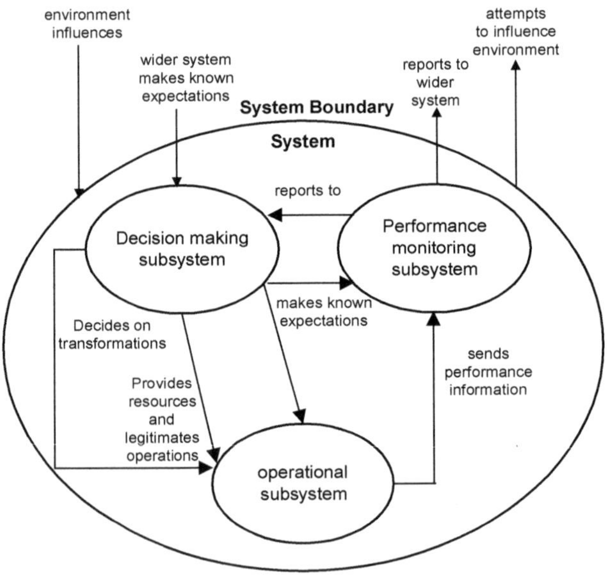

Fig. 5.1 A formal system model of a human activity system (Macauley, 1996)

Although a system places expectations on what the subsystems should do, the subsystems do have a certain degree of autonomy in terms of how these expectations will be realised. This "what–how" relation is a common principle in all systems approaches and is called "hierarchy". Hierarchy should not be confused with authority; it should instead be considered to be a method of "nesting".

The elements of a system in this conceptual model are systems themselves, or (better still) subsystems. Three types of subsystems are distinguished: decision-making, performance-monitoring and operational subsystems.

Unlike the steady-state model, the FSM does not contain the flow of "products" of the system.

5.2.2 The Viable System Model

The viable system model (VSM) of Stafford Beer (1985) consists of functions that must be present in any viable system; these systems all have the same pattern of functions. However, this pattern should not be considered to be an organisational structure. This function pattern forces the researcher to ignore the existing structure, which is usually represented by an organisation chart, enabling the researcher to view the organisation from a different perspective.

The basic concept behind VSM is that in order to be viable, organisations must maximize the freedom of their participants within the practical restrictions of the requirements in order to realise the objectives of these organisations.

Beer distinguishes five functional groups. He calls them systems, because each functional group is a viable system on its own (see Fig. 5.2).

The systems are:

1. System 1: Implementation, consisting of execution, operational management and environment.
2. System 2: Coordination of operational systems. Beer illustrates this using a school timetable. This is a service that ensures that a teacher has only one lecture at one time and that only one class uses a particular room for a given lesson. This service does not affect the content of the lecture—that is under the control of the teacher—but facilitates the operation of the school. A production schedule for a manufacturing utility is another such system.
3. System 3: Internal control. This system preserves the purpose of the organisation "here-and-now".
4. System 4: Intelligence deals with the future environment: "there-and-then". It makes propositions to adapt the purpose. At the organisation level, research departments and marketing typically belong to this group of systems.
5. System 5: Strategy and policy. This system is responsible for the direction of the whole system. "System 5 must ensure that the organisation adapts to the external environment while maintaining an appropriate degree of internal stability." (Jackson 1991).

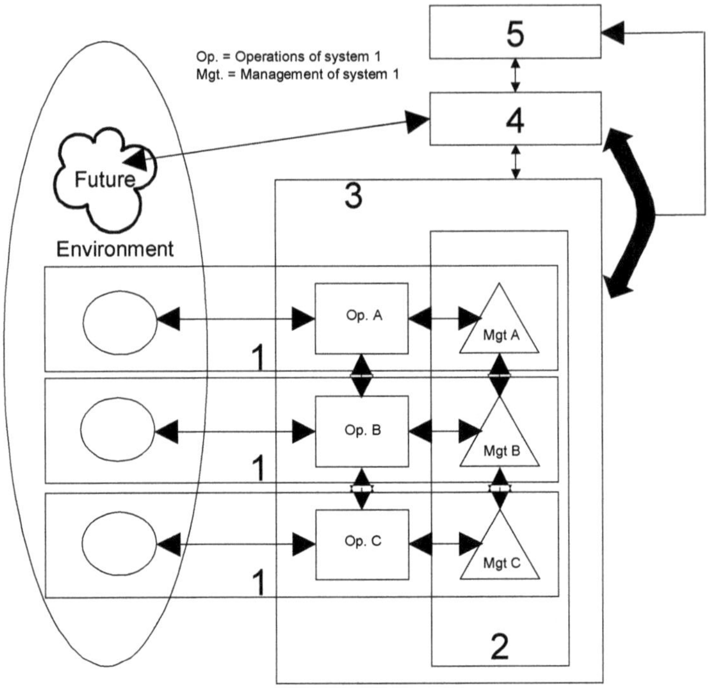

Fig. 5.2 The viable system model (Beer 1985)

Again, the model is recursive; each system is a complete VSM in its own right.

The systems do not necessarily correspond to task descriptions, and several systems can be performed by just one person or department. Just like a FSM, the VSM does not contain the flow of products of the system.

5.3 Common Characteristics of the Conceptual Models

The conceptual models of Chap. 4 and Sect. 5.2 have the following characteristics in common:

1. They are empty with respect to resources and tools. The models specify "what" and "why", but offer complete freedom with respect to physical interpretation. This physical interpretation is exactly the area of each of the domains involved (e.g. organisation, technology and information).
2. All elements are systems and are called subsystems. This similarity of elements is a condition for correct modelling, because there will be no confusion of thought, and zooming in and out is thus defined unambiguously (a system is composed of subsystems, and a set of subsystems is a system). This is the basic principle for being recursive, which is a characteristic of all models. The act of

zooming in can be repeated until the appropriate level of detail for the objective of the research is reached.

3. Every system fulfils a function in its environment by satisfying a need. If the elements of a system are subsystems then the elements also fulfil functions. This "functional approach" is applied to all models.
4. The borderline between system and environment is not defined any further in FSM, VSM and the control paradigm. The steady-state system however, considers the borderline to be a boundary zone that contains functions that fit the flow elements for transformation or delivery by the system.
5. All models clearly distinguish control (decision-making, intervention) functions and operational functions; this is indeed a "paradigm". This paradigm will be used from now on for all system models.
6. We explicitly distinguish "aspects", where an aspect is defined as a subset of all relations. This is expressed by the choice of (horizontally) flowing elements that must be transformed to fulfil the system's function. There is simultaneously a controlling (vertical) information flow within each aspect. This leads to the next statements:
   - The design of a multidisciplinary product requires an approach that takes multiple aspects into account. Several models are required, one for each aspect. There are no strict rules for modelling the coordination and communication between aspects. Each aspect represents a repetitive process and so this coordination and communication will also be a repetitive process with its own standards and interventions. In terms of the VSM, this process is positioned at the level of system 3.
   - Each steady-state model transforms a different flow, one for each aspect, and is controlled by an information flow. The transformation function of this information is the part of the system that transforms the need of the environment (input) into the functional performance (output).
7. Only the steady-state model distinguishes control and transformation functions; the other models do not "produce" anything. The transformation function (input and output zone included) delivers the output (product) by which the function is fulfilled in a physical way. This function is represented in VSM by the operational system (but without flowing elements). The input-throughput-output representation with elements flowing through a transformation function can be applied universally, even to control functions. The physical product there is a control action, standard or procedure. In addition, both services and data (information) can be considered physical products.

The transformation, input and output zones together will, from now on, be referred to as the *process*. The reports from controls to the environment (and higher echelons) will be called the information on functioning or simply the *performance*. Therefore, a system will be modelled by a "PROcess-PERformance" model (a "PROPER" model). Both the control flow and the product interact with the environment (see the incoming and outgoing arrows).

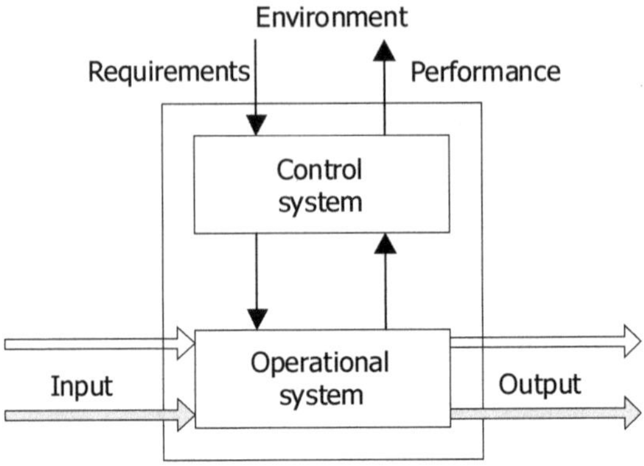

Fig. 5.3 The PROPER model of a system

Upon evaluating the above characteristics, it becomes apparent that a conceptual model of an objective-keeping system must meet the following criteria:

1. The system fulfils a function to satisfy a need of the environment and is purposive. A system will, from now on, be referred to by its function.
2. The elements of a system are subsystems that inherit all of the characteristics of that system.
3. Each system consists of an operational subsystem and a control subsystem; the control subsystem translates a need into the specification for a required product and monitors the operational subsystem; the operational subsystem transforms input into the required output by means of a process.
4. The model distinguishes aspects, which are subsets of the relations. Each aspect represents a product flow. The model of an industrial system includes more than one product flow. The relations between aspects are maintained by a coordinating control system.

Conditions 2 and 3 are quite different. Condition 2 formulates the principle of "zooming"; condition 3 can be best described by "opening" a system at a particular aggregation layer. The aggregation layer just below this one can only be reached by first opening the system and then selecting some of the functions to zoom into. In this way, the principle of control and operation is preserved. For example, the steady-state model is an "open" view of a system (function). One is now able to zoom in on each of the functions of the steady-state model, thereby considering this function to be a system too, but now at a lower aggregation layer.

The resulting conceptual model for an objective-keeping system (the process-performance or PROPER model) is shown in Fig. 5.3.

A function description must be determined through abstraction from physical reality in order to construct a conceptual model. It is not important "how" something is done, but "what" is done and "why". Abstraction to functions offers two advantages:

1. It stimulates us to be creative and to radically change the method of realisation, either through the use of different technological tools and equipment or by combining functions in a different way. It is a structured route to innovation, as opposed to the other path, "accidental creation". The following example illustrates this.

   Example. Damen (2001) reports on a number of attempts at KPN Research that failed to make the mail process more flexible. Finally a research project was started that concluded that only six functions are needed to be able to describe any manner of mail processing: to transport, to store, to pack, to unpack, to merge and to split. Starting from this viewpoint, a completely new and flexible system has now been implemented.

2. If the design is recorded in terms of functions, the basic assumptions and choices made during the design process remain clear and accessible for future design projects. This construction of "memory" prevents the "reinvention of the wheel" and excludes the implicit assumption of superseded conditions.

   Example. During the design process for the highly automated Delta Terminal (in Rotterdam), the organisation of the container "stack" (storage) appeared to be one of the most influential factors on the performance of the terminal as a whole. It was shown that reshuffling containers during their stay in storage would significantly improve the effectiveness of the final transfer process. However, based on experience with manual container handling, the following principle was being implicitly applied: "once stored, a container will not be moved until the moment of final export". The most important reason for this approach was the risk of container loss or damage. However, this risk was not nearly as great with the advent of full automation, and so this principle was no longer valid.

5.4 The "PROPER" Model of Industrial Systems

An industrial system is a subsystem of the organisation as a whole; it contains a subset of the elements, but includes all of the relations. We now approach industrial systems from the viewpoint of the primary function, and at least three aspects are included in the conceptual model:

1. The "product" as a result of a transformation.
2. The flow of orders; without customer orders no products will flow. In this flow, orders are transformed into handled orders.
3. The "resources" (people and means) required to make the product. To make use of them, they must enter the system, and they will leave the system as used resources.

The results of the transformations are delivered products, handled orders and used resources.

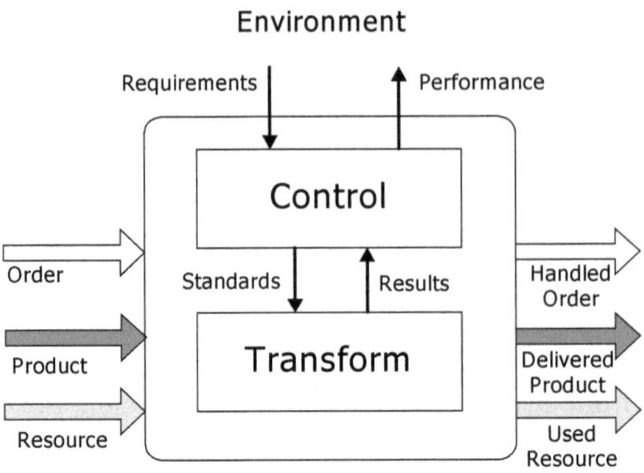

Fig. 5.4 The PROPER model of an industrial system

The PROPER model of an industrial system is represented in Fig. 5.4. The aspects (orders, products and resources) are represented by thick arrows, while thin arrows represent data flows. Usually the flow rate of the resources is slower than the flow rates of the orders and products. People can be employed long-term, and means and equipment will usually be used for their economical lifetimes.

The term "product" does not necessarily only imply a physical product. Especially in the service industry or in the tertiary sector, a product can also be abstract. For example, when the model is applied to an insurance company, the product "an insurance" can be the result of a transformation from a request for insurance.

Information flows both horizontally and vertically. Horizontally, the flow contains (mostly technical) data for the operation itself. Vertically, the flow contains control data.

Within the black box "transform", three parallel transformations can be distinguished:

- The transformation of (customer) orders into handled orders ("perform")
- The transformation of products (raw materials) into delivered products ("operate")
- The transformation of resources into used resources ("use").

The control function coordinates these transformations by generating executable tasks derived from the orders and by assigning usable resources.

Opening the "transform" black box of Fig. 5.4 now results in Fig. 5.5.

Elements of several domains come together in the exchange of tasks and assignments, which is the main focus of communication during the analysis (and design) process. The communication concerns requirements, which are derived from the function's objectives, and feasibility, which is determined by the technological or informational practicability.

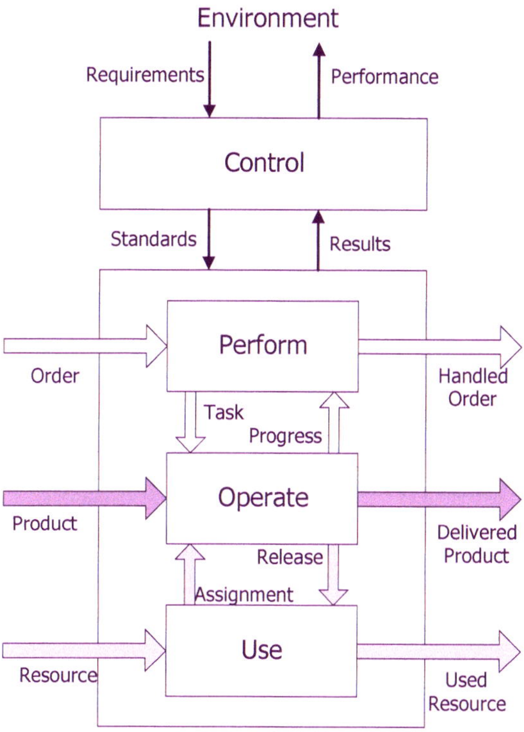

Fig. 5.5 The aspects in the PROPER model of an industrial system

5.5 The "PROPER" Model and Logistic Practice

Until now, our modelling has been generic but quite abstract. It should however be possible to classify the common functions of logistics (a subsystem of an industrial system) that appear in practice into one or more of the chosen aspects.

Logistics is usually divided in material management and physical distribution ("outbound logistics") (see for example Visser and Van Goor 2006). Within material management they distinguish purchase logistics ("inbound logistics") and production logistics. Finally, they also add reverse logistics because of the increasing attention being paid to the flow of returning products. They argue that a logistic concept requires a coherent and iterative way of making decisions about the physical layout (the basic pattern), the corresponding control system, the required information and the matching organisation. The basic pattern refers to the division of basic functions into subsystems. The control system correlates with the control function of Fig. 5.4, while the information required points to both the horizontal and vertical information flows. The matching organisation refers to the organ structure and to the structure of authority and responsibility.

During the last decade material management and physical distribution have been integrated into "supply chain management". In terms of the systems approach, integration means extending the system's boundary and considering the whole as a system that needs to be controlled too.

Towards the end of the 1990s, the Supply Chain Council (SCC) developed a reference model for this integrated approach to logistics: the supply chain operations reference (SCOR) model. This model supports the evaluation of and the improvement of the performance of the supply chain, organisation-wide (SCC 2002). It emerged upon combining business process reengineering (BPR), benchmarking and process measurement. SCOR contains:

- All customer interactions starting from order entry up to paid invoice (see the order flow in Fig. 5.5)
- All material transactions (see the product flow in Fig. 5.5)
- All market interactions, starting with the determination of aggregated needs up to the execution of each separate order (see control in Fig. 5.5)

The model describes four levels of supply-chain management:

- Level 1 contains five elementary management processes: plan, source, make, deliver and return; the objectives of the supply chain are formulated at this level.
- At level 2 the five processes are described more precisely via three process categories: planning, execution and enable. The basic idea is that each of these three categories can be distinguished in each of the processes. The execution category of "source, make and deliver" is further divided into "make-to-stock", "make-to-order" and "engineer-to-order" types. In this way, a complete scheme of 26 possible process categories is created. Any company is able to configure its existing and desired supply chain with this scheme.
- Level 3 shows the information (and software) that are needed to determine feasible objectives for the improved supply chain.
- Finally, level 4 addresses the implementation. Level 4 changes are unique, so specific elements are not defined; only guidelines and best practices are described.

SCOR is a reference model; unlike a conceptual model it classifies all logistics activities, and aims to improve rather than to innovate. All existing configurations can be modelled; solutions that do not currently exist cannot be modelled, as found from the most recent addition of "return".

Plan, source, make, deliver and return are management processes and are part of a control function. To determine which control function they belong to, a short explanation of each is given below.

Plan: Is demand and supply planning and management. It balances resources with requirements and establishes/communicates plans for the whole supply chain. It manages business rules and supply chain performance.

Source: Takes care of the supply of stocked, made-to-order, and engineered-to-order products. It schedules deliveries and receives, verifies and transfers products; it manages inventories, capital assets, incoming products, supplier networks, import/export requirements and supplier agreements.

Make: Concerns the execution of make-to-stock, make-to-order, and engineer-to-order production. It schedules production activities, manages in-process products (WIP), performance, equipment and facilities.

Deliver: Covers order, warehouse, transportation, and installation management for stocked, *made*-to-order, and engineered-to-order products. It includes all order management steps from processing customer inquiries and quotes to routing shipments and selecting carriers. It also includes all warehouse management from receiving and picking up products to loading and shipping products.

Return: Is the return of raw materials (to the supplier) and the receipt of returns of finished goods (from the customer), including defective products and excess products.

If we compare these descriptions with the functions and aspects of the PROPER model of Fig. 5.5, it is clear that:

- There is no strict distinction between aspects in SCOR. Source, make and deliver in particular contain parts of each aspect; in other words, each aspect contains a source, a make and a deliver process.
- The control of the product flow is split between make and deliver. Make takes care of stocks-in-process, while deliver emphasizes warehousing at receipt and shipping.
- Plan contains both the long-term planning and balancing, and the daily coordination of the flows.
- Return represents a complete product flow. In terms of the PROPER model it is an aspect within the product aspect and can be studied separately using the PROPER model.

As argued before, the distinction between aspects is important, because they reflect domain backgrounds and perceptions. It must be clear whether decision-making concerns the order flow, the product flow or the resource flow in order to enable the correct objective to be set. Each flow is controlled by its own control system coordinating the source, make and deliver control functions. Each flow-oriented control system again must be coordinated with the other aspects by a control function at the next higher echelon.

Including the basic processes of source, make and deliver finally yields the PROPER model for each aspect of a logistic system shown in Fig. 5.6.

The definitions of Visser and van Goor (2006) for purchase logistics, production logistics and physical distribution can be mapped one-to-one to source, make and deliver respectively. It is remarkable to see that the field of logistics is divided into functional areas instead of flow-oriented areas. Terms like order logistics, product logistics or resource logistics are not encountered in logistic concepts.

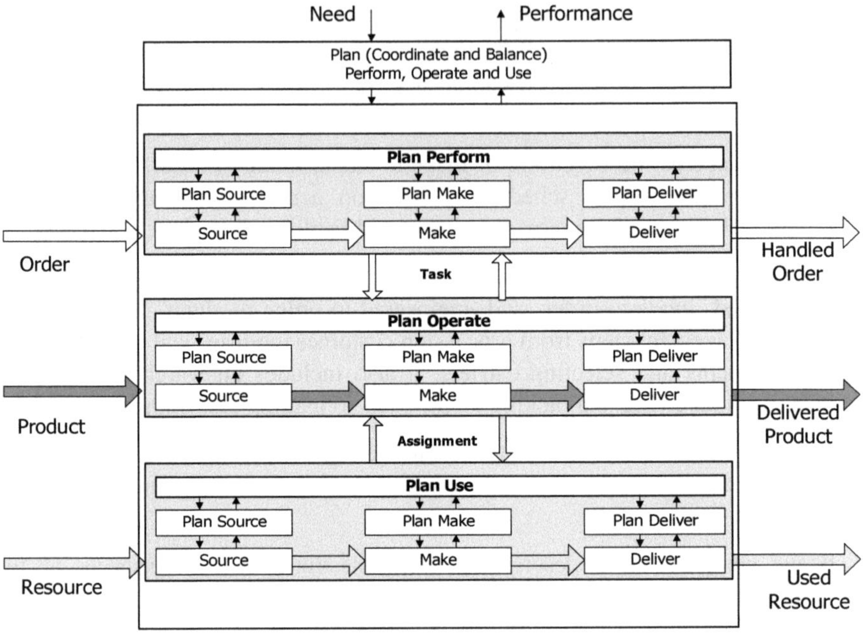

Fig. 5.6 Functions for the aspects in the PROPER model of a logistic system

Well-known concepts like just-in-time and KANBAN can now be positioned in the make process of the product aspect. A control concept like manufacturing requirements planning can be positioned at the level of "plan operate", etc.

References

Beer S (1985) Diagnosing the system for organizations. Wiley, Chichester, UK
Damen JTW (2001) Service controlled agile logistics. Logist Inform Manag 14(3)
Jackson MC (1991) Systems methodology for the management sciences. Plenum, New York
Macaulay LA (1996) Requirements engineering. Springer, London
SCC (2002) The Supply Chain Council (SCC) website. http://www.supply-chain.org
Visser HM, van Goor AR (2006) Logistics: Principles and practice. Stenfert Kroese, Groningen

Chapter 6
Behaviour of a Function: The Process

Abstract. The PROPER model and the steady-state model describe the static structure of a system. In this chapter we add the time-dependent behaviour of a system. Traditionally this has been the field of simulation that is considered to be a real specialism by managers. We are convinced that knowledge of the behaviour is indispensable when analysing or designing an industrial system. We approach communication of the behaviour as a multidisciplinary task. Natural language can be used very effectively to describe the behaviour and specify the simulation model. We show that it should be constructed according the process interaction approach and not according the event description method, which is widely used. Finally, it is shown that behaviour can be described in any phase of analysis and design.

6.1 Introduction

In the previous chapter, the PROcess-PERformance model (PROPER model) was developed and it was shown that it serves as a common reference model for the disciplines involved. It thereby supports designers when dealing with complexity and discussing the analysis of a system. The schematic representation provided by the PROPER model is however static, showing no time dependency at all. As well as analysing the static structure, it is also necessary to study the time-dependent behaviour of the system. This behaviour affects the required dimensions of the system and determines its performance. Therefore, the time-dependent behaviour requires a representation that can be understood and used by all disciplines involved and that is a natural extension to the PROPER model itself. In this chapter such a representation will be presented.

It is currently common to apply simulation models to study the time-dependent behaviour of systems. As with other disciplines, simulation evolved in a monodisciplinary (i.e. predominantly mathematical) culture. In the world of management and technology, this is considered a specialist job. As a result, analysis and design problems are usually "thrown over the wall" to the simulation experts. These experts are expected to understand the specialist requirements, to model the system correctly and to produce accurate results in terms of the problem formulation. In this situation, the quality of the simulation solely depends on the quality of the interpretation of the problem by the simulation expert. The simulation expert, however, is not supposed to be an organisational or technological expert. As shown in Chap. 2, with the systems approach, a "system" is a subjective perception and so a conflict can arise between the problem owner's perception and the simulation expert's perception.

We consider simulation to be a way "to represent system behaviour". In order to represent behaviour correctly, it must first be described in terms that can be understood by the problem owners themselves. Secondly, an unambiguous description should be provided for the simulation expert.

The description of the behaviour of the system under study will be derived from the PROPER model. It will be based on natural language, because natural language facilitates communication and discussion between the different disciplines. Galatescu (2002) states that natural language has unifying abilities, natural extensibility and logic, which are unique properties that enable conceptual model integration. Descriptions of behaviour can be expressed with natural language. In this way, the specification of behaviour becomes the responsibility of the problem owner instead of the solution provider (i.e. the simulation expert). The natural extensibility of natural language also supports the use of conceptual models during the analysis and design process.

In this chapter, the use of natural language to describe behaviour will be examined. When a problem owner explains the way in which a system "works" using natural language, he/she is already describing behaviour globally. The basic idea here is to structure these descriptions in such a way that they can be mapped to the processes of functions in the PROPER model and can still be recognized and understood by the problem owners. In other words, the behaviour of a function is mapped to a "process description" in a cognitive way. The PROPER model is thereby extended with the time-dependent behaviour of the system, preserving the interdisciplinary character. Process descriptions can be defined at each aggregation layer.

A process description is then communicated and discussed with the simulation expert and provides the basis for an unambiguous translation to a software environment, which results in a simulation-specific description. This process-oriented approach has already been implemented in simulation languages like Simula (Sklenar 1997), PROSIM (2005), MUST (1992) and TOMAS (Veeke and Ottjes 2000).

6.2 Behaviour

In Sect. 2.3.3, we explained that

Behaviour is the property of a function that describes the way in which the state of the function together with its input result in output.

Therefore, a function (and thus a system) has two major properties: a state and behaviour.

The state of the function in an industrial system and its input will be defined first. Then a method will be defined to describe its behaviour, thus showing how the state and input together result in output. We will use the terms "system" and "function" interchangeably here. The elements of a system are functions, while the system as a whole is also a function.

6.3 The State and Input of an Industrial Function

The input signals of an industrial function are derived from the PROPER model. Three types can be distinguished:

- *Physical signals*: the order, product and resource flows. These flows might be the outputs of preceding functions (the internal environment) or may enter the function directly from the external environment.
- *Control signals*: these are generated by the control part of a function and passed to the operational part of the function (interventions) or enter the function from higher echelons (requirements). The requirements are translated into standards for the operational function.
- *Disturbance signals*: are parts of the input to the control function (the feed forward and feedback control). They originate from inside the operation part (e.g. machine disturbance) or from the environment of the function (e.g. weather conditions) and influence the progress directly.

If the input signals are represented in the basic industrial function of Fig. 5.3, skipping all output signals, the resulting picture is Fig. 6.1.

Results are the output of an industrial function, while efforts are required to achieve the results. Efforts must be made to provide space, resources, products and time for operation and control. In general, results are the consequence of making efforts to transform the input into output. There are many ways in which efforts can be made to achieve the same results. They depend heavily on the function structure of the industrial system, but also on the relations between the aspects inside the system. Resource assignment and maintenance together with task specification and scheduling influence the efforts being made. By describing the way in which efforts are made, one is in fact describing the relation between input and results: the behaviour. Moreover, there is no need to define the resources precisely; only the important attributes should be defined, such as the speed/acceleration of a vehicle, the processing time of a machine, etc.

Operations research has developed many algorithmic methods for cases that can be formulated in a mathematical way, and often "time" plays no part in these relations. An industrial system, however, is a complex system with many relations and unexpected situations, and each operational transformation takes (or "costs") time; for this type of problem, simulation has proven to be a suitable approach.

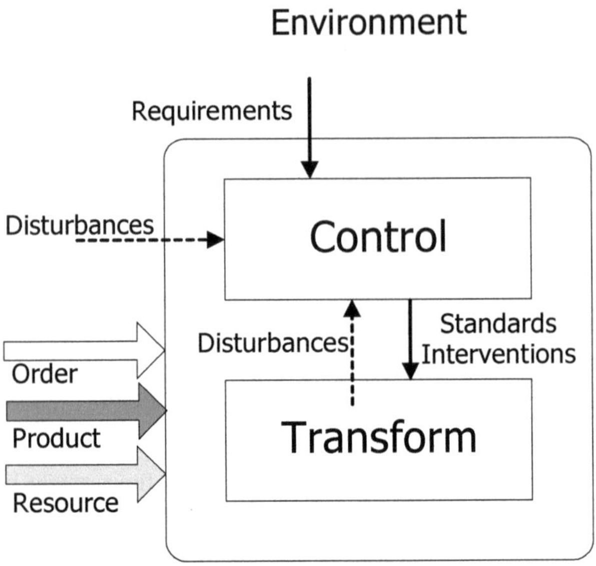

Fig. 6.1 Input signals of an industrial function

Efforts represent the number and duration of resource assignments, the use and number of products and resources, and finally the control efforts. Both the number of effort-requiring elements and the duration of use are important.

The state of a function is defined as a property at some moment in time, and therefore duration must be translated into a value that expresses it at that particular moment in time. According to in 't Veld (2002), the state is the result of input signals and past behaviour, so the duration to be expressed at any moment is the remaining duration. This becomes an expression for the moment in time at which one (or more) value(s) among the set of values will change.

Example 6.1. Suppose a function is "busy" with a task that takes a total of three hours. After one hour of work, a part of the state description of the function will be: "function is *busy* for *the next two hours*".

The example shows that the state of a function not only contains values of static "observable" properties (e.g. busy/idle), but also the period of time during which these properties will keep their current value (e.g. busy for two hours). The properties are derived from the static structure of the PROPER model, while the time periods are derived from the properties of the input signals (e.g. execution times of tasks, working schedules of resources, etc.). These periods must be reflected in the behaviour property.

The PROPER model represents a function structure where each function consists of an operational and control function. The steady-state model (Fig. 4.12) shows all subfunctions in both functions. The model offers a structural way to make an inventory of which functions may be necessary to achieve the desired results. Consequently, defining the properties of each of these functions provides an overview of all internal properties of the overall function. As in the steady-state model, not all properties are required (either because the subfunction is not required or the property itself is not required for modelling purposes); therefore, the required properties will be denoted "significant" properties.

With respect to the behaviour reflecting the periods mentioned, the influence of the environment must be taken into account. The regular part of environmental influence comprises the input flows; beyond these, the function is influenced by the environment in other ways. One may think of weather conditions or unpredictable phenomena that influence the progress of elements through the operational function, such as the moment when a machine disturbance occurs. For this reason, the period will be denoted an "expected" period.

Where the definition of in 't Veld defines the state in relation to input and output (this is in fact an "outside view"), we add a definition that describes how the state is composed (an "inside view"):

The state at some moment in time of an industrial system with a given structure is a set of significant properties of the system, each of which have a value and an expected period of time during which the value holds.

These "expected periods" will constitute the behaviour property of an industrial system.

6.4 The Behaviour of an Industrial System

The behaviour of an industrial system can now be composed in two complementary ways (see Fig. 6.2):

1. As a set of moments in time (also called "events") where the state of the system changes.
2. As a set of periods during which the state of the system remains unchanged.

Both moments in time and periods are determined by the system itself, by other systems and/or by the external environment. When they are determined by the system itself, the term "internal" will be used.

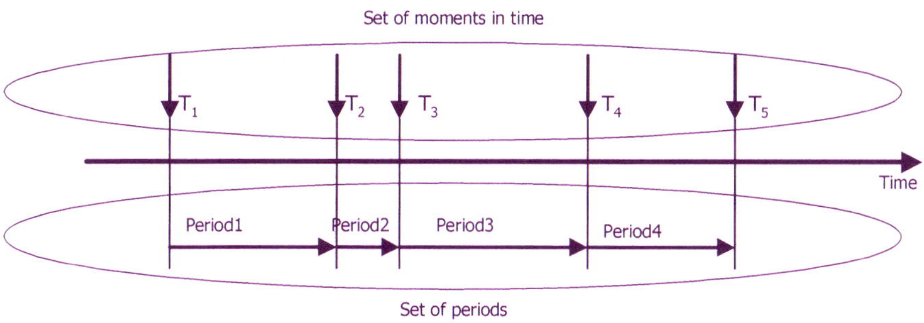

Fig. 6.2 Approaches to behaviour descriptions

In order to decide which representation of behaviour is the most appropriate, the purpose and use of the behaviour description (as an extension of the PROPER model) are first recapitulated.

Firstly, the PROPER model is used as a common reference model for all disciplines involved and thereby supports communication between the disciplines. For this purpose, the PROPER model provides a coherent frame of reference that can be used in an interdisciplinary way. It is, however, a static picture of the system; communication of and decision-making based on behaviour should be supported too. Behaviour and state are added to this concept for the time-dependent description. This description should also be interdisciplinary.

Secondly, it must be easy to structure and detail the behaviour description; during the design process functions are structured in different ways, and zooming in will add details to the descriptions (see Chap. 9).

Finally, it must be possible to relate descriptions at different levels of detail (aggregation layers) to each other. Decision-making proceeds during the analysis and design process, and zooming in to one specific function should not mean that it is isolated from preceding decisions.

Therefore, the major requirements for the behaviour description are: clarity, structural flexibility, ability to detail, and hierarchy.

Behaviour is a complex property and it is better represented by a linguistic description rather than a value. The description explains when a state change occurs and how the system reacts to a state change. The latter part is by definition a part of the behaviour description of the system itself. A state change can be caused by an internal or external cause. Therefore, the behaviour of a system is divided into an internal part and an external part. The internal part contains the way in which the system reacts to a state change and the way in which the system itself causes state changes (for itself or for other systems). *This internal part of the behaviour is now defined as the "process" of the system.*

The external part is represented by the behaviour of other systems. Therefore, the state of a system is not always completely established by its own set of values (including the process), but might be connected to the states of other systems and the environment. This conclusion is completely in line with the principles of the systems approach: "an element cannot be studied in isolation from its environment".

Returning to the two possible ways of describing behaviour, Example 6.2 is used to illustrate both of them and to formulate the arguments for which approach is preferable.

Example 6.2. An operational system transforms semi-finished products from an inventory into finished products. It is assumed that the system is allowed to operate without tasks and that only one resource is permanently assigned to the operation. Therefore, the operation can focus solely on products and the order and resource flow can be skipped.

Products are selected from an inventory in first-in-first-out order; each product transformation takes a certain processing time. If there are no products available the system waits for a new product to arrive in the inventory.

This is already a first process description, although provided completely in natural language. To structure this description, three moments in time are distinguished where the state of the system changes:

1. A product arrives in the inventory
2. The system starts a transformation
3. The system finishes a transformation.

For the transformation there are two periods:

1. A period during which the transformation "works" on a product
2. A period during which the transformation "waits" for a product to arrive in the inventory.

By definition, these periods cannot overlap, so they can be described as one sequence of periods and they must proceed in the order described. Another period must be defined for the elements flowing from the environment into the function through the system boundary:

3. A period between two successive arrivals of a product.

Figure 6.3 shows the moments in time and periods mapped to one time axis.

The terms "work" and "wait" describe the type of "activity" during a period. The generator's periods are apparently all waiting activities for the next product arrival.

As shown in Fig. 6.3, moments in time may coincide. Two sequential period descriptions are required: one for the transformation and one for the generation of products. The next table shows both types of process description using natural language. When interpreting these descriptions, the reader should always be aware of the clock time of the system (which is used by all process descriptions). The current clock time is denoted by "Now".

Each moment in time and each process must belong to the behaviour of some function Operation or Generate Products. This "owner" function is always explicitly mentioned. The same holds for properties. Therefore the property "inventory" is denoted by Operation's inventory. The owner is not mentioned explicitly when the owner refers to its own property. The term "select" here means "select and remove from inventory". Once a period description is started, it runs by repeating successive periods. This repetition is denoted by "repeat".

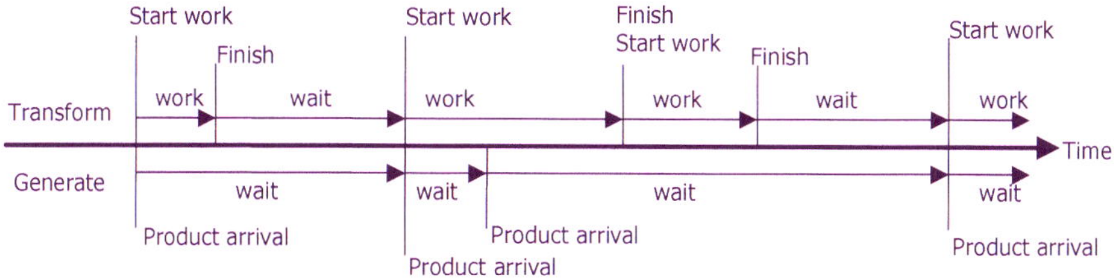

Fig. 6.3 Moments in time and periods for an operational system

Table 6.1 Two ways of describing processes

Processes with moments in time	Processes with periods

Moment_1 of Operation
If inventory is not empty
    Schedule Moment_2 Now
 	
Moment_2 of Operation
Select first product from inventory
Schedule Moment_1 after product's
    execution time
	Periods of Operation
	Repeat
	    Wait while inventory is empty
	    Select first product from inventory
	    Work product's execution time

Moment_1 of Generate Products
Put new product in Operation's inventory
If Operation is idle
    Schedule Moment_2 of Operation Now
Schedule Moment_1 at next arrival
	Periods of Generate Products
	Repeat
	    Put new product in Operation's inventory
	    Wait for next arrival

To schedule the moments in time:
Schedule Moment_1 of Operation Now
Schedule Moment_1 of
    Generate Products Now
	To start the processes:
	Start Periods of Operation Now
	Start Periods of
	    Generate Products Now

To illustrate this interpretation, the first steps of the processing sequence will be explained for both approaches.

Using moments in time:
The clock time is set to zero, so Now = 0. First of all, two moments are "scheduled" to start the processes. Scheduling means that a moment is marked on the clock at the position "Now". Moment_1 of Operation is marked first, so the first line to be interpreted is the one of this moment description. Here the Operation checks the inventory. The inventory is still empty, so Operation does not schedule Moment_2 and becomes "idle"; now the next mark on the clock can be handled. This next mark is Moment_1 of Generate Products. A new product is put in the inventory of Operation. Having done this, Operation must be notified of the value change of one of its significant properties (it is a state change of Operation). However, this value change is only significant if Operation is "idle". If it is already handling a product, then Operation must finish this product first before it can react to this state change. This first time a product is added to the inventory, Operation is indeed idle, so Moment_2 of Operation can be scheduled at this very moment ("Now"). Finally, Generate Products marks its own Moment_1 on the clock so that it will be notified the moment the next product arrives. This moment lies somewhere in the future. The actions of Moment_1 are now finished and the next mark on the clock can be executed; this mark is Moment_2 of Operation, etc.

Using periods:
The clock time is set to zero, so Now = 0. First two period descriptions are "started", one for Operation and one for Generate Products. Both descriptions are marked on the clock at the position "Now". The period description of Operation is marked first, so the first line to be interpreted is the one for this description. It reads "repeat", and this tells the reader to return to this point when the indented lines are passed. The next line says that Operation will wait as long as the inventory is empty. If this command is followed, the condition is "kept in mind" and no mark is added to the clock for Operation. The description cannot be continued because the inventory is empty, and so, returning to the clock, the description of Generate Products is entered since it is the next mark on the clock. After the repeat line, a new product is put in Operation's inventory. This changes the condition "kept in mind" (Operation is watching the inventory), but the description proceeds with "wait for next arrival", which marks this period description somewhere in the future and will then return to the following line (which is "repeat"). Looking again at the clock, a condition is "pending" and one clock mark is available in the future. Before the clock advances to the next mark, the condition is checked. Because there is a product available, the reader now returns to the line in the description of Operation that follows this condition check. The first product is selected and execution starts. This results in another clock mark in the future. At that moment the clock time can be advanced, etc.

The following general conclusions apply (moment in time is denoted by "event"):

1. When using events, there may be descriptions where no change of state appears to occur. In the example, nothing seems to happen at Moment_1 of Operation when the inventory is empty. One could argue that the operational state of the system changes at that event from busy to idle. Although true, this is not clear from the description itself and so it introduces ambiguity. The period approach explicitly states that the system is "waiting" in this case. Periods always express a kind of "activity" or "passivity".
2. The order in which events are described does not influence the course of the process if the first event remains the same. One could exchange the order of Operation's Moment_1 and Moment_2; if the operation still starts with Moment_1 the course of the process stays the same. However, the course must be explicitly mentioned by using "schedule" expressions. When periods are used there's no need for this, because the description order of the periods determines the course of the process. This agrees with the way that natural language is read: from top to bottom.
3. In the description of events, the next event is created by specifying when it occurs. If no next event can be specified (as is the case in Moment_1 of Operation if the inventory is empty), it is impossible to "delay" the current event itself until the next event can be specified. For this reason, Generate products must reschedule Moment_1. The description using periods is able to delay events by extending the current period; e.g. by specifying conditional finish times ("wait while inventory is empty").

4. The sentence "select first product from inventory" can be added to the description of Moment_1 and to the description of Moment_2. There is no reason why it should exclusively be assigned to Moment_2 (because they are the same moments if a product is available); the only condition is that the product must be selected before the function starts executing. This sequence condition is automatically fulfilled when using period descriptions.
5. In the example, the events are named Moment_1 and Moment_2 of Operation. In complex situations, it is advisable to name the events in such a way that they express what actually happens at that moment. The best expression for Moment_1 of Operation would be "the system eventually starts waiting for a product". The period description actually uses this expression ("wait while inventory is empty").

With respect to the first requirement of the descriptions—that they must be clear—all conclusions show that the period approach is preferable so far. Each function also has one and only one description of periods. From now on, a period description will be called a process description.

The example is now extended to include tasks and resources. Then, according to the PROPER model, the behaviour of both a perform system and a use system must be described.

Example 6.3. The operational system of Example 6.2 now requires both a task and a product to start the transformation. Beyond this, the resource must be shared with other operations, so the resource must be explicitly assigned to this operation. It is assumed that each order consists of just one task. A task is added to a task list of the operation and selected in FIFO order.

The processes are shown in Table 6.2.

Looking at the descriptions in Table 6.2, the following general conclusions can be drawn:

1. A number of conditions can be checked sequentially and fulfilled at different moments when using period descriptions. When events are used, sequential stepwise checking introduces an event for each condition.
2. By using period descriptions, the sequence of actions and conditions becomes and stays clear. Interpreting a period description places the reader in the position of the transformation itself. This enriches the problem formulation for the system. For example, here tasks and products are selected at the moment a resource is assigned (upon knowing that at least one task and one product are available). What happens if task and/or product data are needed to assign the resource? Questions like these can easily pop up when confronting each discipline with period descriptions. In this way, decision-making is supported during the design process.
3. Extending the number of functions involves extending the description with one period description for each function (plus one generator eventually). The number of events, however, increases by more than one event for each function. In complex systems the number of events explodes and explaing the behaviour becomes a complex task. Kirkwood (2002) states: "Many people try to explain business performance by showing how one set of events causes another or,

when they study a problem in depth, by showing how a particular set of events is part of a longer term pattern of behaviour. The difficulty with this 'events cause events' orientation is that it doesn't lead to very powerful ways to alter the undesirable performance. This is because you can always find yet another event that caused the one that you thought was the cause".

4. With period descriptions, each system is able to express that it is continuously watching for a condition to become valid by means of "while" and "until" clauses. With events, the condition can only be checked once; other systems must take care of rechecking.
5. If the name of the event description doesn't express the action then the description is difficult to interpret. One carefully note the state of the system. For example, at Moment_1 of Generate Products and Moment_2 of Perform, one cannot simply test for "operation is idle", because the operation can be idle for more than one reason.
6. Combining or splitting systems are straightforward tasks using period descriptions. Suppose the perform system and operation system need to be combined into one organisational unit. In this case, the description of operation can easily be adapted by replacing the first line of its behaviour with the three lines of the perform system. These tasks are not as straightforward when using events. Moment_1 and Moment_2 of Perform cannot simply replace Moment_1 of Operation, because Moment_1 is referred to by other events and all condition checks must be reviewed.

Table 6.2 Two ways of describing processes

Processes with moments in time	Processes with periods
Moment_1 of Operation	**Process of Operation**
If task list is not empty	Repeat
Schedule Moment_2 Now	Wait while task list is empty
	Wait while inventory is empty
Moment_2 of Operation	Enter request list of Use
If inventory is not empty	Wait until resource assigned by Use
Schedule Moment_3 Now	Select first task from task list
	Select first product from inventory
Moment_3 of Operation	Work product's execution time
Enter request list of Use	Remove resource from busy list of Use
If Use is idle	Put resource in idles list of Use
Schedule Moment_2 of Use Now	
Moment_4 of Operation	
Select first task from task list	
Select first product from inventory	
Schedule Moment_5 after product's execution time	

Table 6.2 (continued)

Processes with moments in time	Processes with periods
Moment_5 of Operation Remove resource from busy list of Use Put resource in idles list of Use If Use is idle Schedule Moment_2 of Use Now Schedule Moment_1 Now	

Moment_1 of Generate Products
Put new product in inventory of Operation
If Operation is idle
  And Operation Not in request List of Use
    Schedule Moment_1 of Operation Now
Schedule Moment_1 at next arrival
	Process of Generate Products
	Repeat
	    Put new product in inventory of Operation
	    Wait for next arrival

Moment_1 of Perform
If order list is not empty
    Schedule Moment_2 Now
 	
Moment_2 of Perform
Select first order from order list
Put new task in task list of Operation
If Operation is idle
  And Operation Not in request List of Use
    Schedule Moment_1 of Operation Now
	Process of Perform
	Repeat
	    Wait while order list is empty
	    Select first order from order list
	    Put new task in task list of Operation

Moment_1 of Generate Orders
Put new order in order list of Perform
If Perform is idle
    Schedule Moment_1 of Perform Now
Schedule Moment_1 at next arrival
	Process of Generate Orders
	Repeat
	    Put new order in order list of Perform
	    Wait for next arrival

Moment_1 of Use
Put new resource in idles list
Schedule Moment_2 Now
 	
Moment_2 of Use
If request list is not empty
    Schedule Moment_3 now
 	
Moment_3 of Use
If idles list is not empty
    Schedule Moment_4 now
 	
Moment_4 of Use
Select first resource from idles list
Select first operation from request list
Schedule Moment_4 of Operation Now
Put resource in busy list
	Process of Use
	Put new resources in idles list
	Repeat
	    Wait while request list is empty
	    Wait while idles list is empty
	    Select first resource from idles
	    Select first operation from request list
	    Assign resource to Operation
	Put resource in busy list

From the examples and the reasons given above, we thus conclude that the period descriptions are superior to event descriptions for describing the behaviour of systems. They are clear, brief and, to some extent, unambiguous. Period descriptions are also easily detailed and are flexible when dealing with different structures. For these reasons, only period descriptions will be elaborated further here.

6.5 Basic Concepts of Process Descriptions

In Sect. 6.3 the state of an industrial system was defined as a set of significant properties, each of which has a value and an expected period during which the value holds. The process description used to describe these periods was derived in Sect. 6.4. In this section, the ways in which the properties themselves appear in a process description are investigated and general concepts will be derived. For illustration purposes, the process descriptions in Table 6.2 will be used. The way in which the time durations of periods can be described will then be defined, which will lead to a property that describes the state of the process of a system. The PROPER model is normally used when moving from a global model to a network of detailed models. This process of "aggregation" or hierarchy and its effects on process descriptions will be investigated in Sect. 6.5.3.

6.5.1 Properties

The types of properties of an industrial system are derived from the PROPER model of Fig. 6.5. The first distinction is:

1. Properties of the horizontal flows: order, product and resource (flowing through perform, operate and use)
2. Properties of the assignment and task flows between aspects (between use and operate and between perform and operate)
3. Properties of the control and transform functions themselves
4. Properties of vertical flows (flowing through control).

6.5.1.1 Properties of Horizontal Flows

Horizontal flows represent input elements that will be transformed into desired elements.

Properties of these flows can therefore be delineated into two different types:

- Properties describing the state of the elements in a flow
- Properties describing the state of the flow itself.

Significant properties of the elements are the ones that are changed or used by the transformation and the ones that describe the "position" of the element in the transformation: is it on the input side, is it being transformed, or has it been transformed?

In fact, this means that the flow itself can be divided into three interconnected parts: an input flow, a throughput flow and an output flow. The flow properties therefore reflect collections or sets of elements in each part of the flow. This "set" notion will be shown to be a generic concept over the next few pages. To describe the dynamic behaviour of these sets, the flow's input rate must be defined; this represents the number of elements entering per time unit. This input rate may be the result of a preceding transformation function; in this case no explicit definition is required. If the elements enter the system from the external environment then the input rate must be defined explicitly.

The flow rate of elements passing through or leaving the transformation is the result of the transformation itself, so they are not defined explicitly. They will be measured instead.

This leads to the following general property types.

Flow properties are:
- Input set
- Throughput set (work in progress)
- Output set
- Flow rate: number of elements per time unit entering the input set.

A flow element has:
- Properties that will be changed or used by the transformation
- Properties from set membership.

Referring to the example of Table 6.2, the following properties can be recognized.

The inventory is the input set of the product flow of Operation; the order list is the input set of the order flow of Perform; and no input set of the resource flow of Use is defined. Operation and Perform don't have throughput and output sets defined in their respective flows. The Use function has two sets representing the throughput set: an idle set and a busy set. Each resource can only be in one of these sets. The conjunction of both sets represents the total number of resources in the Use transformation. The throughput set is split to reflect the "state" of the resource: idle or busy. This illustrates the use of set membership to represent state properties of elements.

Flow rates are only indicated globally. Both the order flow and the product flow are generated from the environment; for this purpose two generating functions are introduced. Their process descriptions contain the sentence: "Wait for next arrival". This sentence can be extended to quantify the rate; "next arrival" for example could be replaced by "20 minutes" or a stochastic expression such as "sample from an exponential distribution with an average of 20 minutes".

Tasks are the only elements that have a defined property: the execution time. There are, however, some other properties that are implicitly defined by the process description itself: "an order consists of one task only" and "there is only one resource". Changing these properties will lead to the adaptation of the process description.

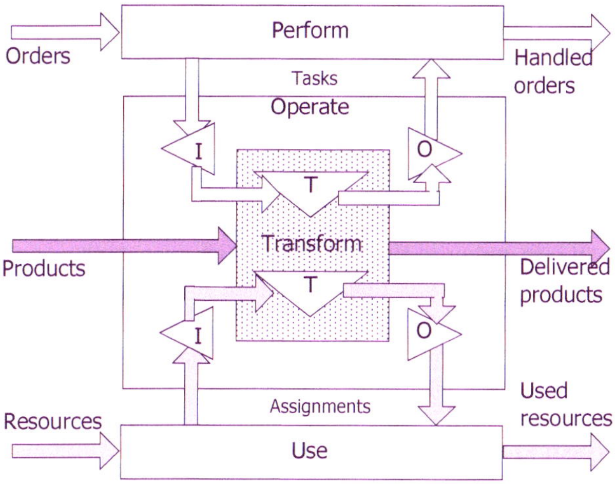

Fig. 6.4 The course of flows between perform, operate and use

6.5.1.2 Properties of Flows Between Aspects

To define the properties of the flows between aspects, the black box of the Operate function is opened as illustrated in Fig. 6.4. In this figure, sets are represented by triangles; "I" stands for the input set, "T" for the throughput set and "O" for the output set.

Task and assignment flows are considered to be input flows of the operation function. In the example of Table 6.2, only one input set has been defined explicitly: the task set. For readability purposes, the input set of resources has been implicitly used by "assign resource to operation" and "wait until resource is assigned". If the input set of resources is called Assignments, then this can easily be rephrased into "put resource in Assignments" and "wait while Assignments is empty".

In general, the properties of these flows are therefore the same as the properties of the horizontal flows.

The concept of sets clearly facilitates the expression of a function's state changes. It even supports decision-making in cases when these sets can contain more than one element, because a selection method is automatically required. In the example all selections are made in FIFO order, but this can easily be changed to more complex strategies with the same concept.

6.5.1.3 Properties of the Control and Transform Functions

Most state properties can be derived from the properties defined so far (flow rates, sets—including their contents) when systems are considered black boxes. The contents of the black box can be represented using the set concept of the flow properties; the dimensions of a system can easily be derived if the entities have properties of space, weight, etc. Measuring the length of stay of elements in a set even results in properties such as lead time, waiting time, etc. The set concept can thus be used to represent the properties of the black box as a whole.

One property, however, has not been considered yet: disturbances. The exact moments of disturbances are by definition unpredictable, but disturbances can be well-represented by a flow and its rate (the well-known mean time between failures, MTBF). These flows are described in analogy with the stochastic flows from the environment. Elements of a disturbance flow are called disturbance; their only special property compared to other flows is that immediate reaction is required when a disturbance "arrives". A disturbance influences the running period of the process. How to deal with this property will be explained Sect. 6.5.2.

6.5.1.4 Properties of Vertical Flows (Flowing Through Control)

Data flow through the control functions vertically. Usually transformation functions are assumed to react based on the data resulting from control. The control function itself may or may not use periods to transform data.

The control function of Fig. 5.5 consists of two connected data handling functions:

- *The standardization function*, consisting of initiation and evaluation.
  - The initiating function:
    - Translates the needs of the environment into standards for all significant properties of the transformation
    - May change the standards in the case of structural deviation of the results of the transformation.
  - The evaluating function:
    - Measures the results of the transformation
    - Reports structural deviations between standards and results to the standardization function.
- *The intervention function,* containing feedback, feed forward and repair of deficiencies.

These both react to individual deviations between measured ("real") values and standard values.

The properties of control functions are often quite specific. The standards must be formulated in terms of the contents of the transformation. Usually this results in standards for the number of elements in sets (representing maximum or minimum inventory levels, number of tasks waiting, available resources, etc.) and the expected residence times in sets (representing desired lead times, accepted waiting times, etc.). The level of detail to which the control function is described depends on the stage of the design process.

There is a clear relation between input, efforts and results. If two of these three factors are known, the third factor can be determined. Assuming that the input of a function is known, there are two different types of goal setting for simulation modelling:

- Determine the efforts needed to achieve the desired results
- Determine the results obtained if these efforts are expended.

The first type is characteristic of decision support during global design; the second is characteristic of decision support during detailed design. Control will not usually be described during global design. Here the goal is to determine the expected values of the properties mentioned using best and worst case analysis and by sensitivity analysis. These expected values can be used as standard values during detailed design.

Control functions will usually evolve during detailed design. This means that it must be possible during the design to add standard values to the sets of a transformation and to the output flows. This requirement will be investigated further in Sect. 6.5.3.

6.5.2 Characteristics of "Periods"

6.5.2.1 Discrete and Continuous Systems

Periods describe intervals in time during which the state of a function does not change with respect to its significant properties. Systems where these "constant" periods occur and the state changes instantaneously are called discrete systems. In continuous systems, the function states change continuously. If a continuous system can be modelled in such a way that the values of its continuous properties are only significant at discrete moments in time, then the behaviour of such a system can be modelled by means of periods. Therefore the distinction between discrete and continuous systems must be considered a modelling criterion; it is not an a priori difference between systems.

6.5.2.2 The State of a Process

It can be concluded from the definition that a period starts after a state change and ends with another state change of the function. The system's clock time advances between these state changes. The system's clock time is the time that all functions are synchronized to.

A state change is represented by a moment in time (event) and a (finite) description of the property values changing at that moment in time. A change in state is required for a period to start. From that point on, it is sufficient to specify the moment in time of the first future state change to fully define the coming period.

If this moment in time is known, the period can be described by "advance the system's clock time until the next state change".

The fact that no state change occurs during the period does not mean that the function is "doing nothing" or "makes no progress". It only expresses that the current state remains unchanged with respect to the significant properties, but the current state can be "working" (for a machining function), "driving" (for a transportation function) or just "waiting" (for a transformation to the next job). Therefore, "work until next state change", "drive until next state change" or "wait until next state change" all express a period (see also Tables 6.1 and 6.2). Indeed, it is one of the major advantages of using periods (with respect to communication between disciplines) that they can be expressed by verbs related to the ongoing activity. We will use the general verb "advance" to express a period.

The other aspect of a period definition concerns the phrase "until next state change". Three situations can be distinguished:

1. The moment in time of the next state change is known exactly.
2. The moment in time of the next state change is not known, but the condition required for it to occur is known.
3. Neither the moment in time nor the condition are known.

Ad 1. If the moment in time is known exactly, the duration of the period is also known exactly, so it is fully defined by "advance x time units". The usual units of time, such as seconds, minutes, etc., can be used here. The only condition is that they must be clear. In this case, the process is said to be in the *scheduled* state.

Ad 2. Here the moment in time is not known, but the condition is known. In this case, an unambiguous expression of this condition is sufficient. A condition can be expressed in two ways:

- The period continues as long as a condition holds; the period is defined by "advance while the condition is satisfied"
- The period ends if a condition is satisfied; the period is now defined by "advance until the condition is satisfied"

In the example of Table 6.2, both alternatives are used. The operation specifies a period with "wait while the task list is empty" and another period with "wait until resource is assigned". Again, the process is said to be in the *scheduled* state.

Ad 3. Both the moment in time and the condition are unknown: the length of the period is indefinite. For these cases, the word "advance" is used without further specification. No state change will occur unless another function (in the surrounding system or the environment) creates one. This can only be awakened by external circumstances. In these cases, the process is said to be in the *suspended* state.

To conclude this paragraph, there is one process state left that needs to be defined. The suspended and scheduled states both reflect periods, but at the instant at which the state changes there is no defined period. Therefore, at the moment that a function handles a state change, the process is termed *active*. A process automatically becomes active at the end of a period in its process description.

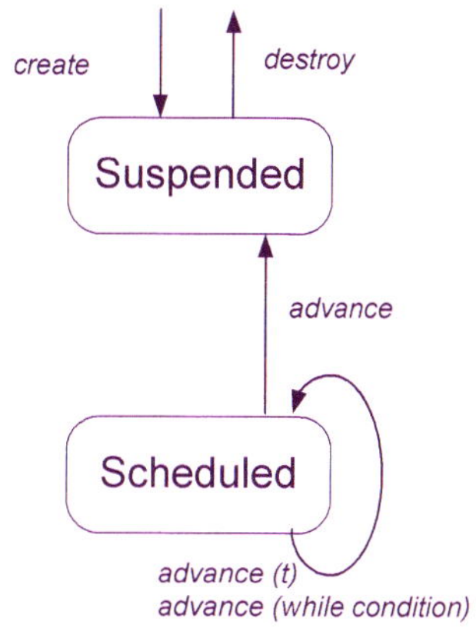

Fig. 6.5 State transitions caused by "advance"

Therefore, the "value" of behaviour consists of two parts: a process description and a process state, which can so far take the values "active", "scheduled" or "suspended".

Figure 6.5 summarises the state transitions of a function's process. These transitions are caused by the function itself via "advance" clauses in its process description. The active state is not shown, because it reflects an instant rather than a period.

6.5.2.3 Process Interventions

If a process is suspended, it will not proceed without intervention from the function's environment. Therefore, facilities are required to intervene with the progress of the process of a function. Interventions can be divided into two categories:

- Regular interventions
- Disturbance interventions.

Regular interventions are best explained using the example in Table 6.2. Consider the next lines from the processes of Operation and Use:

Process of Operation	Process of Use
Enter Request list of Use Wait until resource assigned	Select first Operation from Request list Assign resource to Operation

This part can also be formulated as follows:

Process of Operation	Process of Use
Enter Request list of Use Wait	Select first Operation from Request list Assign resource to Operation Resume Operation

Here the process of Operation proceeds only if another process (in this case the process of Use) tells it to do so. The expression "resume" is introduced to do this.

Resuming a process of a function is the same as resuming the function, because each function can have only one process description. By using resume, the process of operation resumes progress at the line directly after "wait". Resume introduces a new state change from outside the function. The function that causes this must be confident that the receiving function is able to respond to this state change (i.e. it is "suspended"). If the Operation's process is in a scheduled or conditioned state, the period of Operation has already been defined (in other words, Operation is watching the clock time or the condition specified for that period and nothing else). Therefore, Operation would not recognize the state change introduced.

For reasons of readability (and therefore communication), we prefer to minimise the use of indefinite periods (and thus to use formulations like the ones in Table 6.2).

There are, however, two regular interventions that are necessarily caused by the environment of the function:

- The process of a function must be initiated
- There may be circumstances where the process must be stopped.

To start the process at some moment in time, the sentence: "start function at T" is used (the process now changes from a suspended into a scheduled state). As soon as the clock time T is reached, the process will start with its first line and follow its description.

To stop a process, the sentence "stop function at T" is used and the process description is abandoned at clock time T.

Before starting a process and after a process has been stopped, the state of the function is considered suspended.

Disturbance interventions interrupt a process, whatever state it is in, and the function must react immediately to it. The only exception is a process that is active. Such a process cannot be disturbed, because the active state takes place instantaneously: the simulation clock does not advance. As soon as the process description encounters an advance statement, the disturbance is effected.

Usually the reaction of a function will be to do nothing until the disturbance disappears. This is a typical reaction in the case of bad weather conditions or technical disturbances. Of course, repairing disturbances can be a complex task, but one that is usually performed by other functions (maintenance or service). A disturbance will be expressed with "cancel function". The state of the process immediately changes to "cancelled". The process can be resumed after resolving the disturbance in three ways:

- The process must finish the period that it was passing through before the disturbance. The process then returns to the original state. If it was scheduled, the duration of the period will be the remaining period at the moment it was cancelled. In terms of the process description, the process "proceeds with the line it was executing". This result can be accomplished with the sequence:

Cancel function (start of disturbance)

...

Proceed with function (end of disturbance)

In-between, the causing function can describe the actions it performs to deal with the disturbance.

- The process must resume, but the period it was in at the moment the disturbance occurred is skipped. For example, a product that was being handled is removed during the disturbance and transferred to another function. The process may now proceed as if the period is finished and deal with the next state change. Again, in terms of the description, the process resumes with the line immediately following the period before the disturbance. Now the sentences will read:

  Cancel function (start of disturbance)

  ...

  Resume function (end of disturbance)

- Finally, it may be necessary to restart the process completely. In this case, the sentences are:

  Cancel function (start of disturbance)

  ...

  Start function (end of disturbance)

So after a Cancel phrase, a process may proceed, resume or start over.

The process interventions cause transitions in the function's process state. These are shown in Fig. 6.6. This shows that cancel and proceed can be requested for any state of the process that was initiated with an advance statement. A stop request changes the state of the process into suspended. Proceed always returns the process into the state it was in at the moment that it was cancelled. Start and resume always result in the process becoming scheduled.

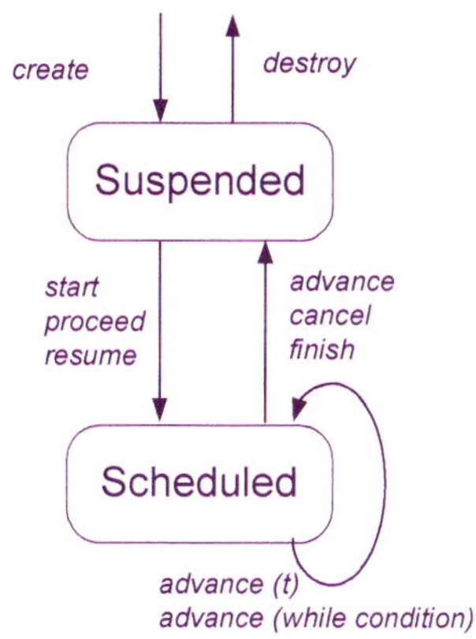

Fig. 6.6 State transitions with an external cause

Start forces the process to execute its description from the first line, while resume forces it to continue with its description from the line immediately following the point where it was cancelled.

6.5.3 Aggregation

Each function is part of a function structure. The static relations are visualised by the PROPER model and the time-dependent relations were described in terms of processes in the preceding paragraphs. During the design process, a hierarchy of PROPER models will be created that represent a number of aggregation layers. Each aggregation layer is based on the results of decision-making in preceding aggregation layers. This shows itself in:

- Refining standards and efforts
- Dividing functions into a structure of subfunctions.

Both actions result in detail being added to the original input signals. In Fig. 6.7 the positions where input signals are transferred between aggregation layers are shown.

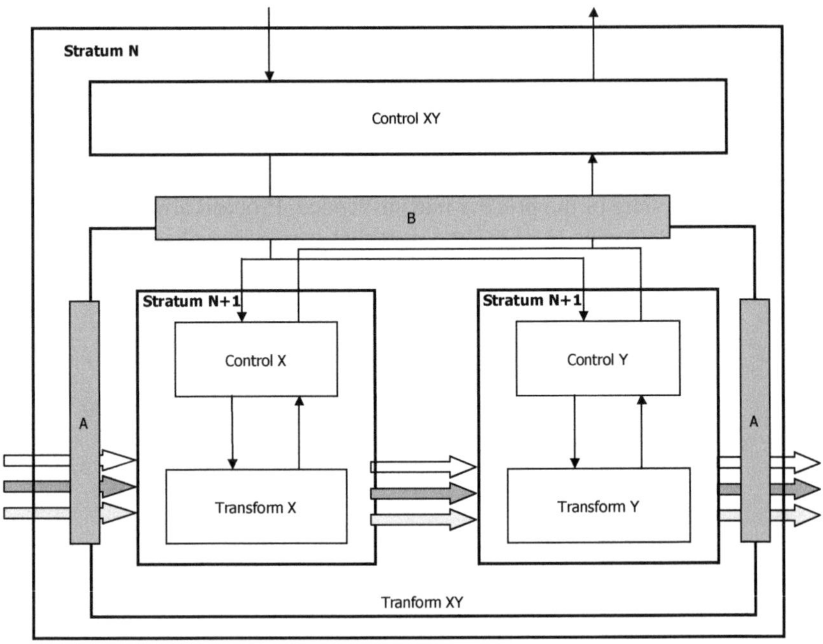

Fig. 6.7 Signals passing between aggregation layers

A function XY is shown at aggregation layer N. Zooming into one step results in two subfunctions X and Y at aggregation layer N+1. There are two kinds of areas where signals are transferred:

- A: for the physical signals of order, product and resource flows
- B: for the control signals.

First focusing on the two areas designated A, an example will again be used to illustrate the effects of aggregation.

Example 6.4. Suppose the transformation XY at layer N represents a painting function. Transformation X at layer N+1 represents cleaning and transformation Y represents colouring. Products can be painted in two colours; 50% of the products must be painted colour A and 50% colour B. If two successive products have different colours then a cleaning action is required first. This takes two hours. The painting operation itself takes five hours. The function model at layer N does not distinguish between cleaning and colouring. The following table shows the process description at each layer as if they were developed completely independently, without any exchange of elements.

Table 6.3 Two aggregation layers compared

Processes at layer N	Processes at layer N+1
Process of Painting Repeat Wait while inventory is empty Select first product from inventory If Sample Uniform[0,1] < 0.5 Work 2 hours Work 5 hours	**Process of Cleaning** Repeat Wait while inventory is empty Select first product from inventory If product's colour differs from last colour Work 2 hours Resume Colouring Wait while Colouring is busy
	Process of Colouring Repeat Work 5 hours Wait
Process of Generate Products Repeat Put new product in Painting's inventory Wait for next arrival	**Process of Generate Products** Repeat Put new product in Cleaning's inventory Determine product's colour Wait for next arrival

The description at layer N can be considered to be part of a model to estimate the painting costs. No specific colour information is needed and it is therefore modelled by the sentence "If Sample…". The descriptions at layer N+1 can be used to determine product selection strategies. The first product is chosen in the example, but this can also be rewritten as "Preferably select a product with the same colour first" to minimise the number of cleaning operations. The main differences between the processes at layer N and N+1 are:

- The function Painting is split into two functions, Cleaning and Colouring
- A product is assigned a property "colour" at layer N+1.

The models can be used in a hierarchical structure; for example, the description of layer N is part of a complete manufacturing model and it is necessary to check the effects of the descriptions at layer N+1 on the results of layer N. In this case, the process description of the Painting function becomes dependent on Cleaning and Colouring. The lead time of the product is now determined at layer N+1, so the sampling phase and resulting periods of two and five hours must be replaced by "Work". The Painting process now simply waits until a signal is received from layer N+1 that a painting operation has finished. The selection of products has also shifted to layer N+1, so this line must be skipped from the description of Painting. The products are still generated at layer N (in this case by a generator, but they may also be the result of preceding functions at this layer), and this cannot be shifted to the next layer. The generator at layer N+1 must be removed.

According to Fig. 6.7, there are two moments in time at which the layers exchange elements. The first one is when an element enters the lower layer and the second one is when it leaves the lower layer (or enters the higher layer). These moments can be positioned precisely and are assumed to be caused by the layer the element leaves. The action of leaving a layer will be expressed by "send element to layer X".

Products must enter layer N+1 at the moment they enter the painting function; this is where they are put into the Painting's inventory. Therefore, the description of the generator must be expanded. Products leave layer N+1 at the moment they are finished by Colouring, so this description will also be expanded. If elements can be sent to a layer, the addressed layer must be able to receive them. To do this, a reception function is introduced at each layer.

This leads to the hierarchically related descriptions shown in Table 6.4.

The process of Painting will now be "working" as long as there are products in its inventory, just as it would be in the description of Table 6.3. Product selection of and operations on them are now completely moved to the next aggregation layer.

The transfer of control signals is represented in Fig. 6.7 by area B. In Sect. 6.3, they were divided into standards and interventions. Transferring standards is a matter of defining new (more detailed) standards. Different aggregation layers are, however, unaware of each other's standards. In Example 6.4, layer N is unaware that cleaning and colouring are separate functions, so it certainly doesn't know what kinds of standards are defined. The same statement holds for measuring results in terms of standards. There is one condition however: the functions must preserve the same behaviour as in the standalone approach. For example, if the occupation of the painting function in Table 6.4 is calculated by measuring the periods where this function "works" or "waits", then this can be achieved in the same way as in Table 6.3. The measured values can be different of course.

Table 6.4 Hierarchically related processes

Processes at layer N	Processes at layer N+1

Process of Painting
Repeat
  Wait while inventory is empty
  Work
 	
Process of Generate Products
Repeat
  Put new product in Painting's inventory
  Send Product to layer N+1
  Wait for next arrival
	Process of Cleaning
	Repeat
	  Wait while inventory is empty
	  Select first product from inventory
	  If product's colour differs from last colour
	    Work 2 hours
	  Resume Colouring
	  Wait while Colouring is busy
	
	Process of Colouring
	Repeat
	  Work 5 hours
	  Send product to layer N
	  Resume Cleaning
	  Wait

Process of Reception_from_N+1
Remove product received from
 Painting's inventory
Cancel Painting
Resume Painting
	Process of Reception_from_N
	  Determine product's colour
	  Put product in Cleaning's inventory

Interventions can be handled in a similar way to the physical signals. Each intervention at aggregation layer N may have consequences at all lower layers. For example, if a control at layer N in Example 6.4 decides to stop the painting function for a while, then this action must be transferred to layer N+1, because cleaning or colouring must also be stopped. This can all be achieved by using the same send and receive concept.

Finally, disturbance signals are considered local to each aggregation layer and these signals do not pass through layers. A disturbance signal is received at one layer and may influence the processes at another layer (up or down), but this must be done by a control function at the receiving layer. For example, a disturbance of the colouring function also interrupts the painting function at the higher layer. This intervention is equal to the "stop" sequence discussed earlier.

6.6 Case: Simulation of the Flight Department

Now let us return to the case of the flight department in Chap. 3.

Veeke (1982) presented the application of process-oriented simulation to this system. From a simulation point of view, this system is rather simple. It however clearly illustrates the differences from a management approach. Where the management approach looks for an acceptable balance between results (lead times, delivery rates) and costs, and therefore adapts the organisation, the simulation approach considers the organisation to be fixed and looks for minimum requirements to cope with stochastic phenomena. In order to illustrate this last approach, we will now describe the behaviour of the system at the level of Sect. 3.3 and then discuss the first experiments performed with the simulation model.

For simulation purposes we will skip the first flight inspection and preparation for delivery. These functions appear to be constant in time, and can be added to the results.

The model will focus on test flights and repair deficiencies. The process descriptions are shown in Table 6.5.

The calculations presented in Sect. 3.3 showed that 18 positions would be required in the hall.

This number of positions was actually provided, but this ignores the term "average". Now there are two options. The first is to organise the flow of airplanes in such a way that an empty position in the hall can be occupied immediately; this option was selected in Sect. 3.3. The other option is to accept the "average" of 18 occupied positions and find a number of positions that would be capable to realise this number.

Using the simulation model, it was shown that the number of occupied positions ranges from 10 to 28. The availability of 28 positions is required to realise a theoretical average number of 18 occupied positions. Based on the simulation model, it was further shown that the availability of 20 positions would be sufficient to approximate the required output of one airplane every 1.5 working days.

Table 6.5 Processes in the flight department

Process of Test_Flights	Process of Solve_Deficiencies

Repeat
 If there are airplanes waiting
  Test all airplanes for 1 day
  For each airplane tested
    Airplane.NrofTestFlights =
    Airplane.NrofTestFlights -1
  If airplane.NrofTestFlights = 0
    Deliver airplane to customer
  Else
    Deliver airplane to Solve_Deficiencies
 Else
  Wait 1 day
	Repeat
	 For all airplanes in hall do
	  Airplane.repairtime =
	    Airplane.repairtime - 1 day
	  If Airplane.Repairtime = 0 Then
	    Deliver airplane to Test_Flights
	  While there are free positions
	   and airplanes waiting
	    Put first airplane on a free position
	  Airplane.repairtime = sample of Fig. 3.2
	 Wait 1 day

Process of Deliver_To_Flight_Department	
Repeat Wait 1.5 day Create new airplane new_airplane.NrofTestFlights = sample from Uniform(3,15) + 1 abort Deliver new_airplane to Test_Flights	

It was shown that the average number of occupied positions would be 17.9, and the average occupation of a position would be 89.5% (= 17.9/20).

From this last result another conclusion can be drawn; the simulation model used an infinite capacity approach to determine the number of positions required. If at a later stage another number of positions is decided on, then it is immediately clear what the utilisation of the positions will be. For example, if 20 positions are selected, then the occupation should be 18/20 = 90%. Because the experiment with 20 positions resulted in less occupation than this (89.5%), the delivery rate will not be 1.5 airplanes per day on average.

The simulation study tackled the issues described in the paragraphs of Chap. 3, such as the influence of bad weather and the required number of radar test cars. We showed the effects of selections made in terms of resources and delivery rate. The simulation showed that the selection of Fig. 3.11 as the organisational approach for dealing with radar and mechanical deficiencies is preferable. It requires at least one radar test car less than the selected option of Fig. 3.12.

In reality, the case took place in the 1960s and, at that time, it was impossible to create a process-oriented computer simulation for it. If, at that time, computer simulation was available and was used in the way described here, the questions could have been discussed together, and management would have been aware of the consequences.

From the example above, it can be concluded that behaviour descriptions (and finally computer simulation) are able to support decision-making, from the very first global modelling steps until the final detailed steps.

6.7 Conclusions

In the previous paragraphs, it was shown that the behaviour of a function can be completely described with a coherent frame of reference, which is based on natural language. It also uses natural language to express decision-making items and offers the flexibility to describe behaviour under any circumstance and to any degree of detail. The number of strictly defined terms in this concept is limited and as easy to adopt as a schematic representation of static function structures such as the PROPER model. Above that, the terms are not bound to a specific discipline and are thus interdisciplinary. As such they support a common description and understanding of the system's behaviour.

Behaviour is a time-dependent property and the notion of time (or periods) is introduced in process descriptions by using time-consuming verbs. The interpretation of process descriptions is straightforward and can be achieved by reading them function-by-function.

For human interpretation the descriptions seem unambiguous. However, in order to be used for simulation purposes they must be unambiguous with respect to automated interpretation. We will not address this issue further here.

References

Galatescu A (2002) Reifying model integration abilities from natural language. J Concept Model 24

Kirkwood CW (2002) System dynamics website on Arizona State University website. http://www.public.asu.edu/~kirkwood/sysdyn/SDRes.htm

MUST (1992) Simulation software user and reference manual. Upward Systems, Rijswijk, The Netherlands

PROSIM BV (2005) PROSIM modelling language (website). http://www.prosimbv.nl

Sklenar J (1997) Introduction to OOP in SIMULA (webpage). http://staff.um.edu.mt/jskl1/talk.html

Veeke HPM (1982) Process simulation as a management tool. Proc. IASTED Int. Symp. Applied Modeling and Simulation, Paris, France, July 1982

Veeke HPM., Ottjes JA (2000) Tomas: Tool for object-oriented modeling and simulation. In: Proceedings of the Advanced Simulation Technology Conference (ASTC 2000). The Society for Computer Simulation International (SCS), Washington, DC, pp 76–81

Chapter 7
The Case of the Ship Engine Factory

Abstract. The use of the PROPER model and the steady-state model are illustrated in this case of a ship engine factory. First, the PROPER model is used to describe the primary processes globally. Then two steady-state models are developed: one for the material flow and one for the order flow. The case concerns an existing organisation where the models are used as a diagnostic aid. They are used to explore the actual course of business and to find the correct problem. All activities in the factory are assigned to the proper functions in two steady-state models. From this questions arise, and they lead to the problem formulation. This case also shows that the models can be used to complete the information required.

7.1 The Models as a Diagnostic Aid

When we design a new organisation with a new objective, we work in a theoretical way by first designing the simplest model and then removing eventual drawbacks by adding functions. We can skip part of this design process by starting immediately with the PROPER model and applying the steady-state model for an aspectsystem. In this way we will design a theoretical "as-it-should-be"-model, to which the future organisation should comply. We thus start from an empty situation and fill it up aided by knowledge of this model. The model is used as a *construction* tool.

In many cases, however, there is an existing organisation, but there are certain complaints about the poor functioning of it as a whole. The organisational expert could of course retire to a room with a group of fellow workers and design a new organisation using the previous method, as if nothing had existed before. Sometimes this may be the preferred method, but we then run the risk of overlooking existing experience and introducing new complaints. In addition, nobody is able to compare the old organisation with the new organisation. Why the existing organisation is failing to function properly remains unclear. It is therefore always necessary to create a *diagnosis* of the facts for the existing organisation before the

H.P.M. Veeke, J.A. Ottjes, G. Lodewijks, *The Delft Systems Approach*,

development of new models. The existing situation, however, is obscure, and it is difficult to determine which functions do not perform well or are even completely absent. In order to take stock of a complex situation, the appropriate method is to create a model. A model is a simplified system at a higher aggregation layer than the system to be studied. When that "as-is" model is ready, we should first of all check up on the agreement of the as-is situation found in practice with and the opinion of the company about what the current situation should be: the *current* "as-should-be" model. For example, we may conclude that the input should be checked and were originally of the opinion that this was being done, but we find from the as-is model that this is not the case at all. Maybe the actual introduction of an input check will solve the stated problem. If the as-is model however corresponds to the current as-it-should-be model, we can, armed with all preceding knowledge, try to recognize the issues with the current as-it-should-be model. After that, we can solve these complaints by improving existing functions, by rearranging existing functions, or by introducing new functions. We then design a *new* as-it-should-be model.

It is more difficult than it seems at first sight for an organisation to set up and analyse an existing as-is situation and to construct the as-is model. However, among all of the preceding knowledge there are some points of condensation, where the knowledge is gathered into theoretical models. We are now able to see the existing as-is situation through the spectacles of those theoretical models. For example: we look through the spectacles of the steady-state model for the diagnosis. In the next section we will describe a particular situation in terms of the steady-state model. By doing this, the reader will learn to *recognise* the theoretical functions. Moreover, the final result of this exercise offers two examples of the application of the steady-state model in practice. The case is based on a real situation in a company.

Section 7.2 describes the technical-organisational course of the business in an engine factory. (Of course, we can also distinguish a commercial aspect, a financial aspect, etc.)

7.2 Description of the Existing Situation

Absolutely no technical knowledge is required to analyse and solve this case.

This ship engine factory constructs rather powerful ship engines. The construction of the engine is such that the power can be adapted to the wishes of the customer by providing the engine with more or fewer cylinders. Usually an engine contains from six to twelve cylinders. Therefore, the production capacity is expressed in terms of number of cylinders per year. This capacity is currently 200 cylinders per year, which is around 25 engines per year. The yearly turnover is 60 million euros. There are 150 operational employees and 80 staff members are engaged.

The main task of the production process is to assemble purchased parts. These parts are partly commodities, and partly parts produced by third parties and engineered on order; the factory has also a small manufacturing department. This department is situated in a separate shop and operates rather autonomously.

Some of the activities of the company consist of offering services and delivering spare parts for these engines. The delivery of spare parts is commercially very important, because it forms a large part of the profit. The slogan of the company can therefore be described as: "Constructing an engine is an investment in the delivery of spare parts".

The assembly is completely organised as a production line. One can distinguish several subdepartments. The following are executed synchronously, and thus in parallel: the pre-assembly of cylinders, the processing of crankshafts, and the start of carter frame welding. The latter will be followed by the boring and assembly of the carter. The final assembly will then assemble these three units into one engine. Finally, the engine is submitted to running tests. These tests take place in a testing station, where three people are working, in the presence of the customer. These three people will solve the detected errors.

Due to the organisation of the production line, engines will sometimes be ready too soon compared to the required delivery time. These engines are then stored in a corner of the assembly shop.

Upon delivery of the engine, extensive user manuals and books for spare parts are also delivered. These manuals are written by a department of five people that adapts them specifically to the engine.

During sales negotiations, the sales department, together with the customer, formulates the requirement specification for the engine to be delivered. They therefore stay in contact with the construction department in order to guarantee the technical feasibility of the engine and to stay within the capabilities of the company.

Sales, however, are running very irregularly. The order book between "sold" and "start assembly" varies from zero to ten engines. Outsourcing the assembly is impossible. If a customer asks for a shorter delivery time than the time calculated from the sequence of order entry, then the only possibility is to change the construction sequence of the different orders. However, once the construction of an engine in the shop has started, the sequence cannot be changed there anymore. A subsequent order can only move forward in the order book between "sold" and "start assembly". The possibilities associated with this are being investigated by the planning department in consultation with the sales department.

Once the requirement specification of an engine has been established, the process planning department (PPD) prepares index cards based on the sales specification, including customer wishes. Using these index cards, the parts that are required for this engine are determined. All parts are identified with article cards, which are selected from a file containing 18,000 different cards. The production line is divided into segments, so each article immediately receives a date on which it should be supplied to assembly. Then the task will move to the materials department, which makes a reservation for the necessary parts on a storage card. Normally, all parts should be in stock because a minimum-stock system is used. The materials department distinguishes three groups of parts:

- A parts, yielding a yearly turnover of more than 20,000 Euros. These parts represent approximately 10% of all parts
- B parts, with a yearly turnover ranging from 2500 to 20,000 Euros, and which represent about 20% of all parts
- C parts, with a yearly turnover of less than 2500 Euros, and which represent the remaining 70% of all parts.

For the C parts, the warehouse works with a so-called two-bin system. One of the bins contains the minimum stock. When this bin is used, a fixed number of parts are ordered. A minimum-stock system is also used for the A and B parts. The amount ordered, however, is based on the demand forecasted for the coming year. In all cases, reaching the minimum-stock value—by reservation or removal of the delivery of spare parts—leads to the ordering of new parts by the materials Department from the purchasing Department. When the part needs to be made in-house, the materials department makes a direct order to the manufacturing department. When the part has been manufactured, it arrives, just like the other externally ordered parts, at the expedition department. An order made to the purchasing department leads to a request for prices and delivery times to potential suppliers. Based on the quotations of suppliers, a supplier is selected. The order is sent to the supplier and a copy is sent to expedition. Storage rules, such as the minimum-stock levels, the number of parts to be ordered, etc., were established five years ago by an external consultancy.

Although the parts should always be in stock at the moment that PPD demands them, there is often a shortage of parts. Lately these shortages have been increasing. Once a shortage of a part is noticed at the moment of reservation by the material department, a warning is sent to the purchasing department. Normally, based on the standard procedures, an order for that part should have already been placed. The purchasing department checks for this, and if an order has not yet been placed, prices and delivery times from suppliers are requested (as mentioned earlier) and a new order is placed. Again, a copy of the order is sent to the expedition department.

Thus, the expedition department receives all parts from the external suppliers as well as parts made in-house by the manufacturing department. Once a part arrives, a copy of the corresponding order is searched for and attached to the part. After this, both the part and the order are shifted to material control, who check the part against the specifications provided by the construction department. If some parts are rejected, material control returns the complete shipment to the supplier. If the quality is acceptable, then the number of parts received is counted. Shortages are immediately reported to the supplier and the received goods are then transferred to the warehouse. Material control communicates directly with suppliers about matters concerning quality. The purchasing department, however, handles communications about shortages and keeps reminding the supplier until everything is received.

The described procedure for looking up parts using index cards is done only four weeks before the planned start of assembly. The parts are removed from the warehouse and stored in intermediate storage, sorted by index card. The job can only be assigned to assembly if all necessary parts are available. Besides delivering parts to assembly, the warehouse also delivers parts to the service department for customers ordering spare parts.

When the assembly orders are issued, they are sent to the foremen of the initial departments. The foreman reads the real starting date from a graph that has been made by planning for this production line. Planning checks progress along the production line based on this graph weekly. Throughput times on this graph are also based on the standard times established by the ergonomist of the company. Both the ergonomist and the planning department watch the feasibility of these standard times and the planned delivery dates continuously. If certain times continue to appear to be unfeasible, new labour studies are performed.

About 200 cylinders are mounted a year. Assuming there are approximately 240 working days per year, this leads to a cycle time of 1.2 days. The complement of workers matches these figures. Every week the progress of cylinder pre-mounting and final mounting is measured and compared to the plan.

Each year, the board of directors determines the number of engines to be produced and a sales forecast for the coming year. This and other things are based on a sales analysis in which the market is screened for the next three years. The board takes also the real production figures from the previous year into account. Were we able to reach the planned production and, if not, what are the reasons for this? All of this information is used to determine an accurate year forecast for the engine production. Eventually, the production capacity is altered to meet the new figures.

The production manager determines the number of cylinders from the number of engines to be produced. The sales manager determines the number of orders to be taken in the coming year and refines this figure for each quarter, taking seasonal influences into account. Both managers follow real production on a quarterly basis.

So far, the description of the situation is taken from a real practical case.

Assignment

Express the description in terms of the PROPER model and describe the functions in the input zone, the transformation and the output zone for both the order flow and the material flow. Describe the function control for the coordination of orders and materials. Then apply the steady-state model to each of the flows. Use the models to state which activities fulfil the functions as described in the steady-state model.

7.3 Solution: Analysis Based Upon the PROPER Model

The PROPER model is used to find the system boundary and the points of interest for further investigation. In this case the resource flow is not an issue.

The first simple PROPER model for this case is shown in Fig. 7.1.

We consider the engine assembly to be the primary function of this company. The system therefore contains all activities that are directly involved in assembly.

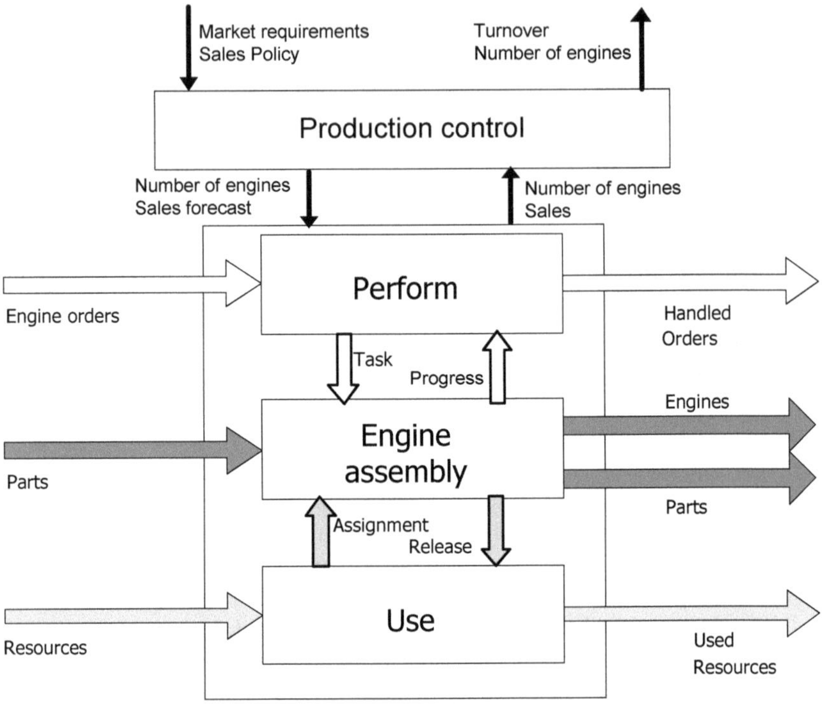

Fig. 7.1 First simple PROPER model

The order flow only contains customer orders for complete engines. The coordinating function Production Control (board of directors) translates (among others) the market requirements for the next three years and the sales policy of the company into standards that can be used by Perform and Engine Assembly. These standards are, for example, the number of engines to be produced during the next year, the sales forecast for the next year, etc. Results will be reported to Production Control in terms of the number of engines actually produced and the real sales, etc.

If we consider the materials for an engine to be the flow elements of the model, then the input is "purchased and self-made parts" and the output "a complete engine and spare parts". The manufacturing department can be considered to be equivalent to a supplier, and this is outside the system boundary.

This model is still very global, so we will have to open up the black boxes Perform and Engine Assembly to get a better grip on the problem. First we will show an open view of the operational functions in these black boxes. After that, each of the black boxes will be separately presented as a steady-state model, with control functions added.

We will combine the descriptions of the operational functions and the control functions and show the steady-state models after that description.

7.4 Solution: Analysis Based Upon the Steady-state Model

Material Flow

The normal course of business in the material flow assumes that all parts for assembly are in stock at the warehouse at the moment that the assembly should start. After all, the company uses a minimum stock system for all parts for this purpose. The warehouse disconnects parts supply from assembly, i.e. it is a buffer function. When we describe this normal course of business in terms of the steady-state model, we see the following functions (each function in the steady-state model is printed in italic):

1. *Encoding*: performs expedition by looking for the copy of the purchase order at the moment that the materials arrive.
2. *Quality control input*: materials department.
3. *Quantity control input*: materials department.
4. *Buffer function input*: warehouse.
5. *Transformation function*: is first of all assigned to engine assembly, but the service department and the customer also use materials (for spare parts). This is a parallel transformation box.
6. *Supporting functions*: no data.
7. *Quality filter output*: this function is fulfilled by running tests in a testing station.
8. *Adding-the-missing*: the testing station where all failing parts and wrong tunings are rectified.
9. *Buffer function output*: the engines waiting for delivery in the corner of the assembly shop.
10. *Decoding*: adding manuals and packing the engine.

These are the operational functions (see Fig. 7.2). Upon adding the control functions we get a complete steady-state model (see Fig. 7.3).

11. *Feed forward on quantity*: The materials department sees that a minimum stock level is reached (compare). If it concerns purchase parts, then the purchasing department takes action. If the parts are self-made, then the materials department takes action (decision). The intervention, the action to be taken, is also executed by the departments. Moreover, the numbers are counted and shortages eventually dunned upon the reception of the order.
12. *Feedback on quantity*: the weekly progress control by the planning department. (Establishing shortages when reserving materials, and subsequently taking action, is feedback on the supply process, and not on the material flow as a whole, and is thus a subprocess.)
13. *Feed forward quality*: The materials department inspects (compares) and eventually returns to the supplier (intervention).
14. *Feed back on quality*: no information.

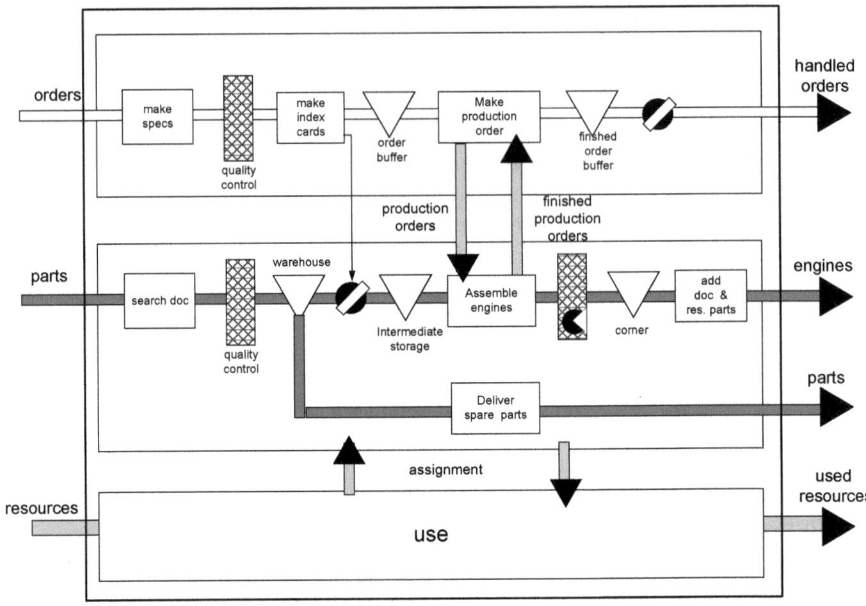

Fig. 7.2 Second simple PROPER model

15. *Initiating functions on quantity*: the external consultancy (five years ago) for the minimum stock levels of the parts, and the production manager for the number of cylinders as he calculates from the number of engines to be produced.
16. *Evaluating functions on quantity*: the production manager fulfils this function for the number of cylinders, and he reports the number of engines to the Board of Directors. This function is apparently missing for the minimum stock levels.
17. *Initiating function on quality*: The construction department, which specifies the purchase parts.
18. *Evaluating function on quality*: no information.

All these functions are represented in Fig. 7.3. The model is drawn as much as possible in accordance with the steady-state model. Of course there are deviations from the theoretical steady-state model. It has already been mentioned that reality will very rarely correspond exactly to this model.

In the model of the material flow, functions have been drawn consistently. For most functions, the department responsible for the fulfilment of the function is shown in the bottom right corner. This enables us to easily survey how the functions are divided between the existing departments, and it may lead to a more logical rearrangement of the functions. On the other hand, when there is no department mentioned in the black box of a function, this function is apparently unfulfilled.

In order to keep the model as clear as possible, we have drawn the control loops for the facet "quantity" above the process flow and the loops for quality under the flow.

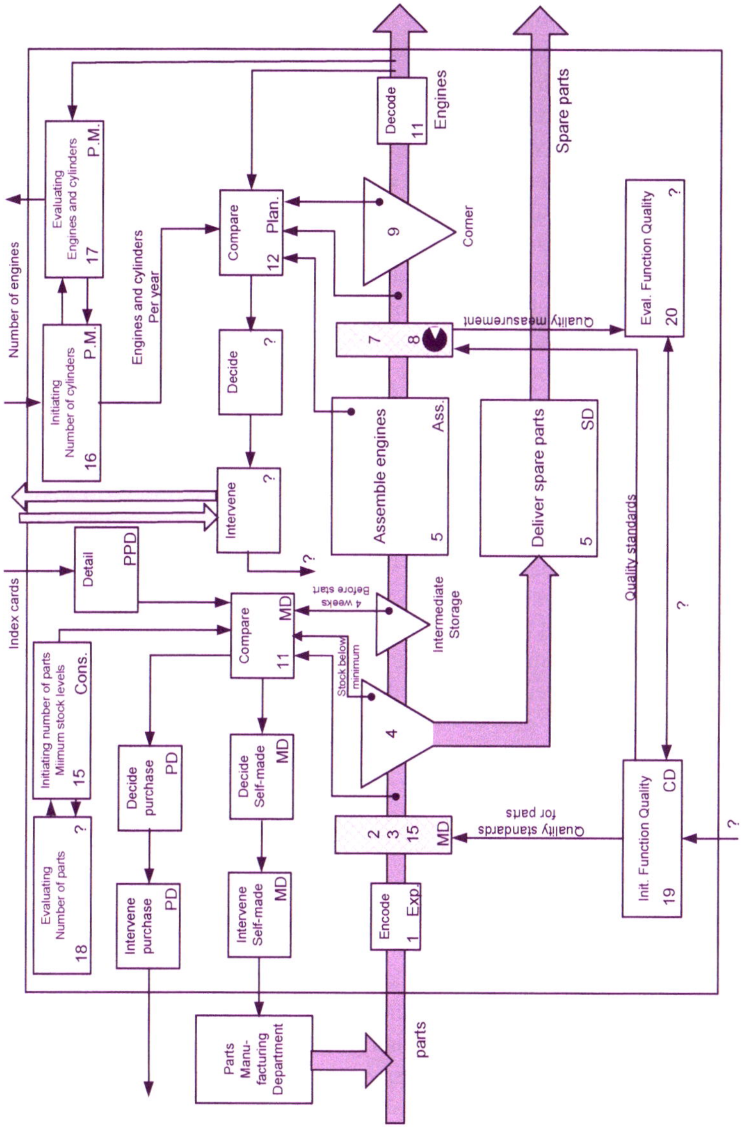

Fig. 7.3 The material flow

Besides this normal process for material supply, where every part should be in stock, there is a separate process for shortages. There are not only measurements of minimum stock levels based on the storage cards at reservation, but there is also a second measurement at four weeks before assembly starts. The available parts are retrieved from the warehouse and placed into intermediate storage. If there are shortages at that moment, the materials department notifies the purchasing department. This is significant, because in a minimum stock system shortages should occur only very rarely if the minimum stock levels are well established. Returning to the model of Fig. 7.3 in order to find a possible cause, we notice that the evaluating function for minimum stock levels is missing. Then, returning to the case description, we suspect that this may be the cause of the growing number of shortages. An external consultancy established these minimum stock levels five years ago, and these levels have never been changed since. The delivery times for parts have increased considerably during these five years and the number of cylinders produced annually has also increased. It is therefore no wonder that shortages are occurring more often. The company reacted by performing measurements four weeks before the start of assembly and trying to receive the parts on time through rush orders and special actions. This reaction was wrong. A symptomatic treatment was applied instead of looking for the cause. The consequence of this was running around in circles; the number of shortages kept increasing under these circumstances. The only real solution is to calculate all minimum stock levels again and to order parts based on these levels. It will take some months before the problem disappears, considering the delivery times, but this is the only way to really master the problem. The Board of Directors made an error five years ago in that it forgot its own evaluating function. This is indeed an error we are coming across increasingly often. Different cases show that the tasks of a consultancy are not implemented in the organisation after the departure of the consultancy. Sometimes one could blame the consultancy for this.

This case is an example where the as-is situation corresponds to the current as-should-be model, but where a new as-should-be model should be constructed in order to solve the real problem. The Board of Directors however did not want to change the current as-is situation into a new as-is situation, in accordance with the new as-should-be model. Our models do not provide a definite approach to conquering this type of resistance, even when the results of the model are rationally convincing. For this we need the support of other disciplines.

Based on the case description, it was impossible to give a decisive answer about various functions due to a shortage of data. In reality we would of course then gather these data. In this way, the models are helpful for checking the completeness of the investigation. In this case for example, quality control has been examined insufficiently. It would be very interesting to investigate the evaluation of quality standards based on customer experiences with the engine.

Of course, some function may be missing in the real situation and it may not be necessary. We should make ourselves sure of this very thoroughly.

Especially in the material flow, other opinions are possible when we choose anther system boundary or zoom into one aggregation layer or choose another flowing element; for example, “the shortage signal”.

In Figs. 7.3 and 7.4 we maintained the terms of the steady-state model in the black boxes of the functions in order to maximise the recognizability. In a practical application we would have focused much more on formulating the functions using terms that are used in the company.

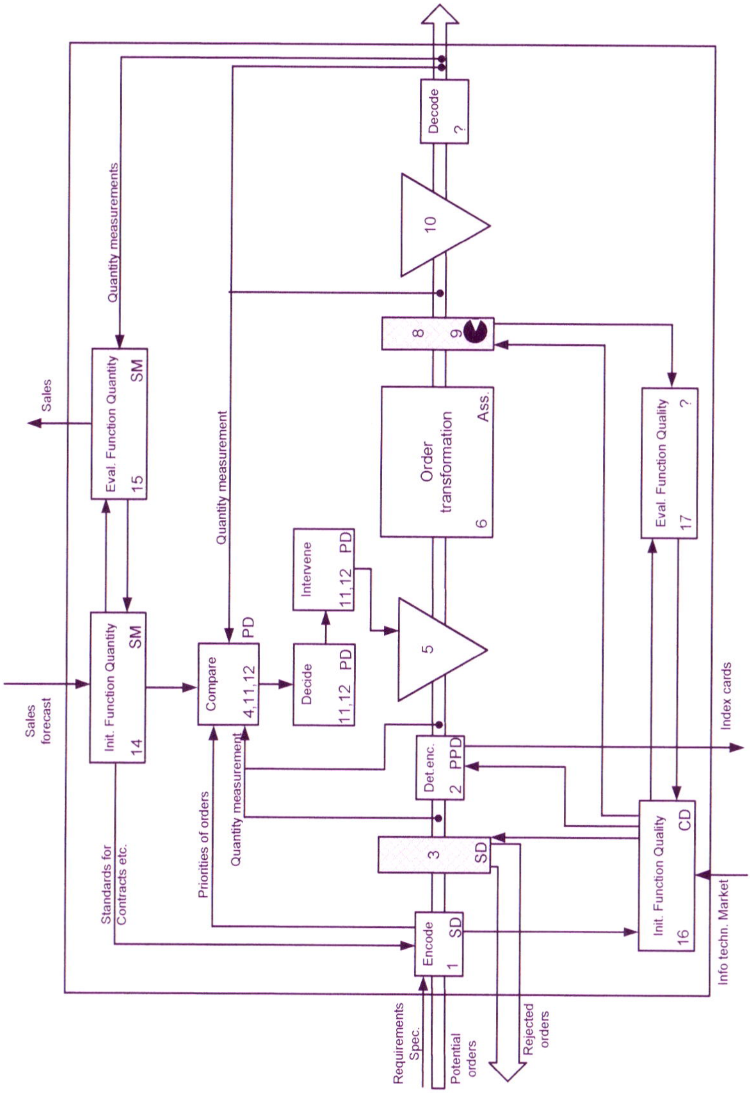

Fig. 7.4 The order flow

The Order Flow

If we consider the order for an engine to be the flow element, the input of the process will be "potential orders" and the output will be "handled orders". The next list specifies where or by whom each function in the steady-state model (printed in italic) is fulfilled.

1. *Coding* is performed by the sales department by drawing up a requirements specification.
2. After this, the order is coded again when a definite order has been received. This is done by the process planning department by making index cards.
3. *Quality control input* is handled by the sales department in consultation with the construction department regarding technical feasibility.
4. *Quantity control input* is done by the planning department in consultation with the sales department.
5. *Buffer function input*: orders that are not yet in progress. The sequence of production for these orders can still be changed by the planning department.
6. *Transformation function.* This is the material production of the order in which it is transformed into a handled order.
7. *Supporting functions*: no data.
8. *Quality filter output*: checking the order from the customer and the company during test runs.
9. *Adding-the-missing*: during quality control output.
10. *Buffer function output*: for orders that are ready too soon.
11. *Feed forward quantity.* The functions "compare", "decide" and "intervene" are fulfilled by the planning department. The intervention consists of changing the order sequence in the buffer function input; during the process there may be interventions during production.
12. *Feedback quantity.* These functions are also fulfilled by the planning department.
13. *Feed forward and feedback on quality.* Information is not available. Maybe these loops do not even exist.
14. *Initiating function quantity*: The sales manager translates the yearly sales forecast into quarterly figures (sales, orders).
15. *Evaluating function quantity:* The sales manager, also dependent on future expectations.
16. *Initiating function quality*: not very clear, eventually the construction and sales departments.
17. *Evaluating function quality*: very unclear. Seems to be missing.

The order flow is modelled in Fig. 7.4.

Both the order flow and the material flow have been considered to be at the same aggregation layer here. The next step in our analysis would be to zoom into each function black box at each step. We are planning, for example, to study the purchase function in depth. The main question then becomes: what happens to the signals in the purchase department that come from the materials department? In this process the steady-state model can again be used as a standard due to the "nurse's effect".

Chapter 8
Policy and Performance

Abstract. If we consider an industrial system to be a structure of functions, a complete series of goals for the system will be found. However, we are not able to assign any weights to those goals or to express opinions about the way in which we think that those goals can be realised. When analysing an existing system, the weight between goals can be found from the structure itself and from extra information obtained from interviews with the management. When designing a new system, the situation is different. If we include these matters too early in the discussions then the differences in opinion quickly escalate, all sorts of interests begin to play a role, and emotions often run so high that a communal, open discussion of *all possible* goals goes completely out of the window. This is the main reason why we attempt to clearly separate the determination of the goals from the formulation of policy, in contrast to many writers who consider the determination of goals to be a part of the policy-forming process. But just what do we mean by "policy"? It appears that productivity, effectiveness, efficiency and performance should be defined unambiguously.

8.1 What is Policy?

It does not appear to be very easy to find an unambiguous definition of "policy". The dictionary states: 1. management; 2. consultation, deliberation, circumspection. If we then look under "management", the dictionary gives: policy. We must therefore look for another approach.

When participants in an organisation say "there is no policy around here", they often mean that:

- Management has different ways of dealing with similar cases; that is to say, management is not consistent—it does not follow a particular line

- The direction in which management is heading is not obvious or is simply unclear; that is to say, there is no clear direction or course—no known end-goal
- There is a goal, but the manner in which management intends to reach that goal, or the methods used to do so, are not known.

In order to determine a particular policy, one must have at least one goal or at any rate a direction in which one wants to proceed. In order to realise those goals, the company needs to have at its disposal:

- People and means (resources) —the system's content; within these we can further distinguish *active resources* which can be used over and over again and *passive resources* that are used up
- Organisational structures that bring people and means together in mutual relationships; those relationships are particularly manifested in the stream of information and communication processes used for decision-making
- Inputs of temporary elements.

The system can, to a certain extent, influence these three factors, and does just that in order to realise its goals. Different structures, however, can lead to the same end-behaviour and thus to the same goal (the principle of structural indeterminacy). Consequently, there are still all sorts of options.

Policy therefore involves the selection of ways and means (an organisational structure is also a means of production) that are used to realise the goal. But this is still not enough to define "policy". There are ways and means that are regarded as "indecent". When determining the policy, as well as legal conditions, unwritten rules concerning general norms, values, or principles are also taken into account. These "principles" are certain fundamental principles, starting points or rules that are taken as standards for our behaviour. Furthermore, we sometimes hear the expression: "that's not the way *we* do things". Therefore, in addition to the general norms and values, we also employ our own specific principles. For example, some companies wish to invest large sums of money in production sites but not in imposing lobbies or in large offices, according to the motto "that is not the norm around here". Management takes these principles into account when choosing from the possible ways and means. These principles are less susceptible to change than the ways and means. However, they are usually not formulated in advance. They are obvious from the decisions taken in practice as a trend in those decisions becomes clear.

Policy is then the selection of the ways (the manner), the people and the means that should be employed to realise the goal. Policy-making is, by and large, a twentieth century discovery.

Indeed, it took until around 1900, after the transition from handicraft to industrial production and under Taylor's influence and the advent of scientific management, before it was realised that work must be analysed and prepared in a more scientific manner. Process planning and execution are separated at the lowest level, the factory floor. In 1916, with the appearance of Fayol's book (1984), a better, thought-out approach to total management come to the forefront. However, it was only after 1960 that large-scale research in this area was begun and attempts were made to develop a theory. Corporate strategy forms part of that.

If there is only one goal, this determines the content of the concept "policy". However, organisations always have more goals. What's more, two situations need to be distinguished:

- The goals are agreed upon but insufficient means are available for realising all of the goals concurrently. Priorities must be set. The goals are *competing*.
- Two goals formulated by different stakeholders cannot both be realised. One cannot simultaneously turn left and turn right. The goals clash; they are *conflicting*.

Usually both cases occur in a series of organisational goals. In both cases a political choice must be made. The results of this choice are strongly determined by the balance of power and by the weight that each group assigns that goal within its own prioritisation of goals. It is usually a question of bargaining and compromises. Note that a compromise does not necessarily have to be bad. It is a compromise in the negative sense of that word when viewed through the eyes of each individual stakeholder. However, seen from a larger system of which the organisation is a small part, that same compromise can be the best solution for that larger system. In some organisational views, it is assumed that individual and organisational goals must coincide in order to achieve an optimally functioning organisation. However, in each organisation some individual's goals will be in direct conflict with those of the organisation. We should not turn a blind eye to this. Conflicts of interests between individuals and the organisation in which they work exist, and will always do so, whatever the political constellation. A benefit for an individual is usually a sacrifice for the organisation (e.g. salary versus labour costs). However, one must constantly strive to obtain a satisfactory balance.

Despite their differences, all stakeholders have one common interest: the survival of the organisation. The primary goal of the organisation cannot be determined autonomously by that organisation. This places limitations on the stakeholders' positions of power and on the possibility that the choice between certain conflicting goals will lean too heavily to one side. In practice, it is sometimes the case that the parties concerned treat the choice between two goals as if those goals are conflicting, whilst further consideration shows them to be simply competing. This means that the selection problem has been approached and discussed incorrectly. We should continuously ask ourselves whether the goals do indeed exclude each other or whether they are only competing goals.

In the case of competing goals, those goals can, in principle, all be realised if there are enough resources available. However, if the resources are limited, one must decide which goal will be afforded priority. One then often sees that the maximisation of one goal is pursued. This, however, automatically results in the sub-optimal achievement of all other goals. Even if we prioritise the goals, so that they may never degenerate into the pursuance of the absolute maximisation of the most important goal, then the realisation of all remaining goals is pushed too forcefully into the background. Also, a decent balance must be sought between the competing goals when priorities have been assigned.

Norms and values also play an important role when determining the priorities of competing goals. In numerous organisations, economical–technical values such as profitability, productivity and efficiency are central. This is something that, through education and experience, is embedded in our subconscious such that we are hardly aware of this when making decisions. Whilst it is often said that other values were also considered during discussions, it appears from the decisions made that economic criteria dominate over and over again. Therefore, for example, in recent years and in the shadow of American companies, the values of the shareholders and stock prices receive amazingly high priorities.

In our opinion, it is extremely important that we are aware of this so that we can ask ourselves whether the financial–economic dimension is being overemphasiised, causing a healthy balance of priorities to be lost. In difficult situations, the question remains "which values do we sacrifice and to what extent, and where do we draw the line?" Thus, values also play a large role in policy. However, those values will change over time; including amongst the organisation's management. The management and the employees are not immune to political and cultural changes in the environment of the organisation. After all, they are also part of that environment.

Just as we did with the system's and the subsystem's goals, we must also distinguish different levels in policy, such as division, sector, department and work group. At each level the objectives of that unit must be determined. They are, however, subgoals of the goal at the level just above this one. The system forms the environment of the subsystem. The unit must then decide on the ways and means (policy) used to realise the goals of that unit. This policy must remain within the framework decided by the higher level. Policy-making is not just the task of the Board of Directors; a line manager must also determine a policy for and with his group. However, the policy of that group encompasses a smaller area, is far more detailed and must remain within the framework decided by the Board of Directors.

One extremely important aspect, namely the timescale, is still missing. In the policy, when do we want each step towards the goal to be realised? This timescale is added to the policy through scheduling. The schedule also breaks the path to the goal into steps and coordinates these steps.

To summarise: in order to be able to determine policy we must first and foremost have *goals* or at least a *direction*. Policy then encompasses:

ways (the methods used)
+
people and means → and policy principles
+
priorities of the goals

Scheduling adds the timescale to the policy, elucidates the path to the goal in terms of steps, and coordinates these steps.

For goals, policy and scheduling, we ought to distinguish between different levels. The goals, policy and scheduling at a lower hierarchical level must remain within the framework provided by the goals, policies and schedules of higher levels. Policy therefore provides a structured framework within which the decisions made by everyone in the organisation fit continuously.

This composite of goals, policy and scheduling should be dynamic rather than static. Almost every category is in a continuous state of flux due to exchange between the system and the environment. The system's outputs cause changes in the environment that in turn can cause changes within the system. The policy must aim at flexibility; in other words it should *anticipate* expected changes.

8.2 Does an Industrial System Need Policy?

Some Boards of Directors are rather opportunistic. They often think in the old trading mindset and are of the opinion that freedom in decision-making is a necessary ingredient for success. This is short-term thinking that does not inspire confidence among other groups of stakeholders. What's more, policy is necessary so that everyone involved can contribute like-mindedly to the same goal. The content of that policy must then be clear to all. A clear policy also enables the lower levels to become more independent in terms of decision-making. *It is only possible to delegate functions and authority when there is a clear policy in place.* The system's procedures can be described in less detail. When everyone is familiar with the goals, ways and means, *better reactions, characterised by more flexibility and more speed, are enabled to changes in the situation at all levels in the company's hierarchy.* The company, as a whole, becomes more decisive. At all levels, policy-forming is a slow-moving process of becoming aware that must be consistently enforced. Greater decisiveness is the result. Policy is formed and carried out not just at the top level but at all levels, by the sector, the department and the group. This must remain within the framework laid down by the policy at the top, but it is policy nevertheless. When we try to formulate the policy briefly and concisely on paper it appears to be a strong stimulus for raising awareness. Of extreme importance are discussions about the draught with and amongst the employees. As a result, all sorts of aspects which the draughtsmen have not thought of will come to the fore. This collective thinking-through and development of the policy brings it to life for all of those involved, and so they care about it more. In addition, it makes general announcements about the policy easier and fewer misunderstandings are possible.

Policy deliberations and formulation are strongly influenced by the company's history, its traditions and its culture. From the literature, we sometimes get the impression that we automatically arrive at a policy using the methods of corporate strategy. However, in principle, these methods provide no more than a thorough analysis and preparation after which the final decisions about which policy will be followed (based on many other issues too) will be taken by the *entrepreneur*. Entrepreneur and entrepreneurship are not synonymous with manager and management skills. Entrepreneurship also takes into account the behaviour of the system. It expounds and also encompasses the taking of risks in relation to innovations in products and organisation, and with it the acceptance of both risk responsibility and liability. While some managers are entrepreneurs, many more are good managers but don't dare to take risks and are thus not entrepreneurs.

It is very detrimental when the policy appears to change frequently. It then appears that the *preparative thinking* has then been too shallow.

In particular, over longer periods, we must ask ourselves whether:

- There will be any other interested parties in x years time
- Any of the interested parties (old and/or new) in x years time will perhaps have other wishes than the current interested parties
- We expect society's norms and values to have changed in x years time;
- The norms and values of the employees will change; for example, in terms of the acceptance of the present authority relationships in the organisation.

On the other hand, the policy should not be a dogma. The policy must be adjusted when there is a change in the starting points on which it is based. When the policy is well thought-out and concisely formulated it is easier and faster to work out why it must be changed and which points need changing. We can thus arrive more quickly at a new standpoint. The most important point however is that we are aware of why the policy is changing and what the consequences of doing this are. However, we should point out that the employees and other directly concerned parties must be ready for policy changes. A certain policy is only possible with a certain internal situation and in a certain environment. *If necessary, tailored training courses or enlightenment courses are required in order to achieve this.*

However, we must not work out the policy in such great detail that it encourages opposition to a subsequent policy change. Rather, the purpose of policy formulation is to bestow direction and to determine a course of action. In short, to create a certain stability in direction, not in detail. Stability should never degenerate into rigidity.

Policy is necessary in many areas. Each goal requires a policy for each of its aspects, for example:

Product policy
Production policy
Quality policy
Market policy
Investment policy
Research policy
Policy for the image of the organisation
Personnel policy
Career policy

Organisational structure policy
Standards policy
Policy-forming policy

8.3 Considerations When Choosing the Ways and Means

Activities must be undertaken in order to achieve the goals. Something or someone must carry out the transactions. For this, a person or a means is required—a machine, a tool, the reader, someone else. In order to acquire that means or to get that means to carry out the transaction, the system and that person must make sacrifices: monetary sacrifices when buying the machine, energy sacrifices that allow it to function, sacrifices in terms of limiting the freedom of the individual, etc. The term "sacrifice" is used intentionally here rather than "costs". We can only talk of costs if the sacrifice can be expressed in monetary terms. When achieving goals in social systems, however, other sacrifices play a role that cannot be expressed in terms of money; sacrifices such as working against one's will or in uncomfortable circumstances, with limited opportunities for self-development. The term "sacrifice" encompasses all experiences that are less than pleasurable or are a burden, which are thus more than just physical effort. Free time and job satisfaction are generally experienced as yield, just like the salary. Effort, the work itself and a limitation on the freedom of the worker imposed by authority are often seen as sacrifices made by the worker. Determining what constitutes a sacrifice is absolutely subjective. One person considers "working" a sacrifice, whilst another person considers it a yield (can and must work!).

Both the yields (results) and the sacrifices are thus subjectively determined values. With both concepts we must acknowledge not only the quantitative but also the qualitative aspects. The commonly intended result must mean an acceptable sacrifice for each person involved, even if different workers weigh up the pros and cons differently. In the concepts that follow on from here we must continually bear in mind that for both the sacrifices and the results, it is the general content of those concepts that is being referred to. Our starting point is that each person and each system strives to achieve the highest possible yield for the least possible sacrifices in the broadest sense. Some people "work" (sacrifice) as little as possible in order to "do nothing" (yield) afterwards for as long as possible. That is also striving to achieve a large yield from the sacrifice, to achieve high productivity.

In the discussion of system goals, it was suggested that the goal is the desired result from the activities to be performed. However, there is usually a choice of different means for performing these activities. The process of selecting from among those alternative ways and means is the *strategy* (the broad-line determination of and selection from alternative means). For each means we can theoretically determine what the expected result will be upon applying that means. The actual result of employing that means only materialises after the choice has been made and the means is actually used. This actual result can deviate from the theoretical result on which the choice for the mean was originally based. Three sorts of results can thus be distinguished:

1. *Intended* result
2. Theoretical *expected* result
3. *Actual* result.

Whether the intended result is necessary for survival, or simply desirable, is not considered here. In both cases, the intended result is the intended consequence of a resolute and purposeful transaction. However, in the first case, the demand to achieve the goal weighs more heavily than in the second case.

The term "result" is used here. Both positive and negative values can however be assigned to one result. The term "yield" would actually be better here. However, for practical reasons with respect to the following formulae, "result" (R) in the sense of "yield" is used, in contrast to the term "sacrifices" (S).

In terms of the sacrifices, we can distinguish between the theoretically expected sacrifice from applying a certain means and the actual sacrifice incurred after the application, in other words between:

- *Expected* sacrifice, and
- *Actual* sacrifice.

So there are two types of sacrifice and three different results. In an organisation we can hardly pitch the concept "intended sacrifice" opposite "intended result". The sacrifice that we want to offer is always as small as possible, whereby the intended result must of course be achieved. Therefore, the concept of *maximum permissible* sacrifice can be useful. In this case we cannot concurrently place a demand for an intended result. We must wait and see what the highest result is that we can achieve for that maximum permissible sacrifice. If we start out from an intended result then we could attempt to minimise the sacrifice. However, if the minimum sacrifice to be offered up in order to achieve that goal is actually greater than we can afford, then we must abandon the intended result.

When choosing from alternative means it is first and foremost important to gauge whether a certain means is effective; that is to say, whether it can enable us to achieve the intended result. In the case of a compound goal, the degree in which we realise that complex entity of partial goals is important. The effectiveness concept is applicable here.

Effectiveness is defined as the relationship between two results (*R*/*R*). For the various means we determine the *theoretical* effectiveness based on the results expected by the supplier based on theory when applying this particular means:

$$Effectiveness_{theoretical} = \frac{R_{expected}}{R_{intended}}. \tag{8.1}$$

When the theoretical effectiveness of a certain means appears to be less than 1 then we are no longer interested in that means. This indicates that we cannot use it to reach the intended result. However, when this is the case for all of the means examined, then we must either look for other means or we must accept that the intended result cannot be realised using known means. The only remaining possibility is then to set a more modest goal. A means can therefore fall short ("undershoot", have an effectiveness of less than 1). However, a means can also do more, or even something other, than what is required to achieve the set goal ("overshoot", have an effectiveness of greater than 1). This occurs, for example, if we also include the time in which the goal must be achieved in the goal formulation. For example, say that the goal is to travel to Paris from Amsterdam. A plane, train, car, bicycle and walking are all effective means. If, however, we set as goal of travelling to Paris from Amsterdam within six hours, then walking and cycling are eliminated from our considerations. However, the plane gives an overshoot because it enables us to arrive in Paris well within the timeframe permitted. Nevertheless, a lathe that can make 1000 products per day exhibits *over-capacity* when only 500 products are required per day. However, we could ask, "what is the certainty that that lathe could indeed make 1000 products per day, or that the plane reaches Paris within 6 hours?" These figures are based on specifications from the lathe manufacturer and the airline, based on normal conditions. But what will the performance be in *our* situation? All sorts of unexpected difficulties could crop up; for example, it could be a foggy day when we choose to travel. Particularly in complex situations, we must also take into consideration the *probability* that we can achieve the goal with a certain means. In other words, how sure are we that, in a certain situation, the means will indeed do what we think it will do? We cannot always oversee all facets. Trying it out can increase our insight, but this is not always representative or even possible. In Sect. 8.5 you will find an application of these concepts to the example of a "plague of rats".

Yet another aspect emerges when a certain goal needs to be achieved repeatedly. We then have to take into account the *precision* of the means. That is, the expected spread of the results around the intended goal; for example, the spread in quality. Over-capacity, probability and precision will often play a role in the definitive choice of the means to be applied. In addition, we must know how likely it will be that the means will be out of order when we want to use it. This factor of *reliability* is also important. When choosing the equipment for a space capsule, the reliability is generally far more important than when the same equipment is applied in a terrestrial machine. We must therefore look at and weigh up all these factors in view of the *environment*: the conditions in which the means will have to function. With all of these considerations, it does not make any difference whether it is about an economical goal or about a concrete or abstract goal or means.

Following these comparisons of the theoretical effectiveness of each of the various means and after weighing up the pros and cons of other factors, a number of means will be dropped. But how then do we make a choice from the remaining possibilities? To do this we must also examine the sacrifices that we must provide for and with those means.

Productivity is the relationship between result and sacrifice (R/S).

The ultimate choice must be made based on the *theoretical* productivity of each of the remaining means.

$$Productivity_{theoretical} = \frac{R_{intended}}{S_{expected\ with\ that\ means}}. \tag{8.2}$$

Here, both the results and the sacrifices are still the supplier's written specifications of the theoretically expected values. After all, we do not have those means in-house and so we still do not know what the actual results and sacrifices will be in our situation.

We then choose the means that, on paper, demands the least sacrifice; in other words, the means with the highest theoretical productivity, at least when the goal is absolute and definite. If we suspect that a higher intended result may be necessary in the future, then we already take that into account now. In that case, "the maximum productivity of that means" is a useful concept.

$$Productivity_{theor.max.} = \frac{R_{max.\ expected\ with\ that\ means}}{S_{expected\ with\ that\ means}}. \tag{8.3}$$

We therefore differentiate between the current and the future situations and accept a lower than maximum productivity for now.

If two means have the same effectiveness and productivity but one is directly available and the other must be fetched from somewhere, we choose the first one of course. In some cases this can be so important that we still prefer the directly available means even if this results in a somewhat lower productivity. This *availability* must also be taken into account when weighing up the choices.

In many practical situations it is a very good idea to perform a sort of sensitivity analysis. Sometimes it appears that the sacrifices are reduced significantly if we accept a slightly reduced effectiveness or formulate the goal slightly differently. Sometimes a little more sacrifice results in a much greater increase in effectiveness. In such cases we must reconsider the starting points: the goals set.

Summary

First comes the question, "can we actually realise the intended goal with existing knowledge, means and methods?" The very first question therefore concerns the effectiveness. Only after that does the question of productivity arise.

In social systems, the goals are realised through concerted action between people and means. It is then often useful to (and provides a better insight when we) differentiate between:

- The productivity of the person
- The productivity of the means
- The productivity of the whole; in other words, of the combination of people and means (in 't Veld 1993).

For the first and third of these types of productivity, we can only express some of the sacrifices and yields in monetary terms. Usually we can reduce the sacrifices offered by the person by offering greater sacrifices for better means. In other words: when we invest more money/more capital in better means then we can avoid human sacrifices such as the need to work in an unhealthy or tiring work situation. What is a yield for the individual here is often a sacrifice for the system. We must therefore consider yields and sacrifices from two different standpoints.

8.4 The Concepts of Productivity, Efficiency and Performance

During policy-making, we must select between alternative ways and means. In the last section we saw that the ways and means must first and foremost be effective. As the expected results could only be calculated on paper, this was defined as the theoretical effectiveness in Eq. 8.1.

In order to be able to choose from the resources that have a theoretical effectiveness that is equal to or greater than 1, the concept of theoretical productivity was also introduced, in Eq. 8.2.

The resource chosen, at the *strategic* level, on the basis of these considerations has now been acquired. The chosen resource is introduced into the system. We are now going to use the resource in practice. We are only concerned with one resource, the chosen one. Often the same resource can still be employed in different ways. The task of selecting from among these different ways of applying one resource is the *tactic*: the detailed determination of the alternative employment possibilities of one given resource and the selection of the most appropriate one.

Firstly, we must determine what the best way is. This is the method that realises the maximum attainable result using that resource with the lowest sacrifices, for our situation. We also set this *maximally attainable productivity* to be the *standard* that we must attempt to realise in practice. So:

$$P_{standard} = \frac{R_{maximally\ attainable\ with\ that\ means\ in\ that\ situation}}{S_{minimally\ attainable\ with\ that\ means\ in\ that\ situation}} = \frac{R_{standard}}{S_{standard}}. \tag{8.4}$$

The standard is thus determined at the tactical level for the productivity to be realised at the *operational* level. In reality, it is often not possible to do this due to all sorts of disturbances. Both the results and the sacrifices could differ from the set standard. That is why the following is implemented at operational level:

$$P_{actual} = \frac{R_{actual}}{S_{actual}}. \tag{8.5}$$

Naturally, at this operational level, we regularly compare this actual productivity with the standard productivity determined at the tactical level, in order to approach the latter as closely as possible. The difference between the two P values could be due to a difference between the results, a difference between the sacrifices, or difference between both. It is impossible to determine the cause of the deviation from that one ratio between the two productivities, those on the result side and on the sacrifice side. In order to make this possible, we perform two separate steps.

1. Assume that the *sacrifices* are *the same.* The actual sacrifice is the same as the minimum sacrifice, and thus the same as the standard. We now divide P_{actual} by $P_{standard}$, and so the same sacrifice appears in both the numerator and the denominator and they cancel each other out. We then get:

$$\frac{P_{actual}}{P_{standard}} = \frac{\dfrac{R_{actual}}{S_{actual}}}{\dfrac{R_{standard}}{S_{standard}}} = \frac{R_{actual}}{R_{standard}} = Effectiveness_{actual} \,. \tag{8.6}$$

When the standard is actually reached $R_{actual} = R_{standard}$, and so the actual effectiveness is equal to 1 (100%).

2. We could however also assume that the *results* are *the same*. The actual result is the same as the maximum result attainable with that resource in that situation, the same as the standard. Once again we divide both productivities by each other. Then the same result is present in both numerator and denominator and they cancel each other out. This gives:

$$\frac{P_{actual}}{P_{standard}} = \frac{\dfrac{R_{actual}}{S_{actual}}}{\dfrac{R_{standard}}{S_{standard}}} = \frac{S_{actual}}{\dfrac{1}{S_{standard}}} = Efficiency \,. \tag{8.7}$$

We call this ratio of two sacrifices the *efficiency.* In contrast to the concepts of "productivity" and "effectiveness", the concept of "efficiency" can only occur at the operational level. The normative efficiency is therefore 1 (or 100%). More detailed study of the method and practical experience indicate that S_{actual} can sometimes become even smaller than the $S_{standard}$ that was set. The efficiency is then greater than 1. We have therefore found a new $S_{minimum}$ and thus a new standard. After the tactical level has distributed these *new* standards in the system, the efficiency number then drops back to 1. In practice, instead of the meaningless ratio between

P_{actual} and $P_{standard}$, at the operational level we control:

- The actual effectiveness against standard 1
- The (actual) efficiency against standard 1.

We therefore compare the results and sacrifices separately. It holds that:

$$P_{actual} = \frac{R_{actual}}{S_{actual}} \,.$$

In any case, we can multiply this twice by the number 1:

$$P_{actual} = \frac{R_{actual}}{S_{actual}} \times \frac{R_{standard}}{R_{standard}} \times \frac{S_{standard}}{S_{standard}} .$$

If we move the individual factors around here, we get:

$$P_{actual} = \frac{R_{standard}}{S_{standard}} \times \frac{R_{actual}}{R_{standard}} \times \frac{S_{standard}}{S_{actual}} ,$$

and this means:

$$P_{actual} = P_{standard} \times Effectiveness_{actual} \times Efficiency . \qquad (8.8)$$

At the operational level, we employ the resource both resolutely and purposefully. If we assume that the goal would indeed be achieved, then the only question remaining here is whether the resource is used efficiently. At the strategic policy level, the question of primary concern is:can the resource be effective for the goal we have in mind? At the tactical level, it is about the most productive employment method for the given resource in that situation. That method is then set as the standard at the operational level. During actual daily use at that operational level it is about the actual effectiveness and the efficiency.

The question at the *operational level* is: how do I reach or exceed the set standards? At the *tactical* level, the question is: how can we put the given, available resource to its best possible use? How can we best perform a given task? At the *strategic level*, the questions are concerned with the theoretical effectiveness and the productivity of the yet-to-be-chosen resources.

It is important to realise that a given organisation can operate absolutely efficiently and can spend an awful lot of time on this without the organisation being effective. Hijmans refers to this as "aimless efficiency". In the first place, this is about asking whether the correct task is being executed or whether it is possible to completely avoid that task. Only then does the question of whether the given task is being executed correctly become interesting.

In conversation and in the literature, there is some confusion about these concepts. Indeed, it appears that all disciplines understand "effectiveness" to be the ratio of two results and "productivity" to be the ratio of result to sacrifice. The use of the word "efficiency" (S/S), however, changes that situation.

Examples:

- The Encyclopaedia for Business Economy defines "efficiency" as the ratio between the input and the output.

  That is: $\frac{S_{actual}}{R_{actual}}$.

- Van den Berg provides the following definition: "The efficiency of an organisation is the measure wherein the given organisation's goals are achieved using a minimum of organisation's resources and/or the measure wherein given resources contribute maximally to the realisation of the organisation's goals." This is actually *R/S*, the productivity. This definition of efficiency comes from economics.

- Eilon, however, defines:
  Machine time efficiency = output of product A / max. possible output of A.
  This is *R/R*, the effectiveness.
- On the other hand, Van der Schroef says: "Efficiency is a comparison of the costs, calculated on the basis of the standard, with the actual sacrifices spent." This is *S/S*, the efficiency as defined above.
- In public administration, in 't Veld Sr. defines "efficiency" as being a desired goal that is achieved with the least possible resources, or the highest possible effect being obtained with the given resources.
  This is again *R/S*, the productivity.
- Gooding and Wagner differentiate between absolute values such as net income produced and relative values such as the output:input ratio. They then call all the absolute values of achievement "productivity" and all relative values "measures of efficiency". This also stems from public administration.
- From sociology, Etzioni defines: "An organisation's efficiency is measured on the magnitude of the resources that are used to realise one production unit."
  That is to say, *R/S*, the productivity.

Therefore, in many cases, efficiency and productivity are used as synonyms for the same type of ratio, namely, *R/S*. The person who distinguishes between both words sees a system as being more efficient when it has a greater productivity. Therefore, the efficiency is a ratio between two productivity numbers. This latter definition approaches the deduced definition because in most cases we compare two systems with the same goal. The results (the goals) are then the same, and the greater productivity is simply the result of a lower sacrifice. Indeed, in this case, the greater productivity *R/S* is caused by a better efficiency, *S/S*.

This confusion of terms is less of a hindrance when we are aware of the three ratios *R/R*, *R/S* and *S/S* and their mutual coherence. We can then test each definition against them.

Up until the end of the nineteenth century, "effectiveness" and "efficiency" were regarded as being almost synonymous. The Oxford Dictionary suggests, for example, that efficiency is the fitness or power to accomplish the purpose intended; effectiveness, efficacy (*R/R*). However, around 1900, efficiency acquired a second meaning: the ratio of input to output, such as effort to result, expenditure to income, costs to resultant satisfaction (*R/S*). In scientific management, a third meaning has emerged: efficiency is the ratio between the costs of the actual execution and the costs of the standard execution (*S/S*).

We must take care not to confuse results and sacrifices in practical applications. Let's say that the standard time required to make a product is 60 man-minutes (1 man-hour). However, one particular worker does it in 50 minutes. We may then be inclined to say that the result from this worker is 50 minutes instead

Fig. 8.1 "Excellent work Jansen. It no longer leaks, it looks professional and it took a long time to complete … That's the way I like it."

of the expected sacrifice of 60 minutes. However, both are and remain sacrifices. The result is one product. In this case:

$$Effectiveness_{actual} = 1$$

$$Productivity_{standard} = \frac{1 \text{ product}}{60 \text{ man minutes}}$$

$$Efficiency = \frac{60 \text{ man minutes}}{50 \text{ man minutes}} = 1.20$$

Sometimes, it is even more complicated than this. Such a situation can arise when a sacrifice for one person (the paying customer) is a result for another (the plumber who cashes in). Figure 8.1 illustrates this.

These concepts conjure up a very "economic" impression.

That is true if we see things as Hennipman (1962) views them: "The common core of the issues with which economic science is concerned lies in the relationship between relatively scarce, alternative useable goods and the entirety of needs or purposes from whatever sort, for which the goods are useful".

When we also see goods and purposes as immaterial, then issues such as work climate, people's happiness, personal development, etc., can be interpreted as "results" and "sacrifices". Also, these concepts must test the processes that are designed for such goals. However, this means that we can no longer lump different factors (such as money) together, and that the value that we assign to something is purely subjective.

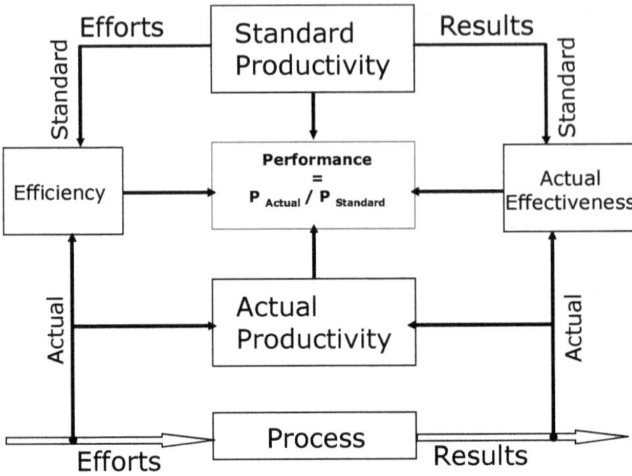

Fig. 8.2 The coherence of the concepts

We could also show the coherence of the three concepts in another manner. To carry out a process, sacrifices must be fed in. We could say that sacrifices are imported (an input). After the process, the results come out of the black box; see Fig. 8.2.

These results produce a flow of money that runs in the opposite direction to the process flow and largely serves to compensate for those sacrifices. We should continuously control both the imported sacrifices and the realised results and compare these with the standards that have been set. The *performance* delivered is then the ratio of the actual productivity to the productivity set as being standard. According to the formula that was derived for this, then:

$$Performance = Effectiveness_{actual} \times Efficiency\,. \tag{8.9}$$

In this way, we control the performance of the process over shorter and longer periods of time. When we discover a deviation between the standard and the actual productivity over a short period of time (the performance is insufficient), we will attempt to trace the reasons for this and take action to achieve the standard.

The split into effectiveness and efficiency gives us an insight into whether we should look for the causes on the sacrifices side, on the results side, or on both. This is more or less analogous to the daily control loop in the steady-state model. If we control over a longer period of time, then the *average* effectiveness and the *average* efficiency are compared with the standards. If there are deviations that we could not bring under control in spite of all of the interventions made, then that can be a reason to change the standard for productivity after evaluation. This is then more analogous to the standards control loop with initiating and evaluating functions from the steady-state model. The model from Fig. 8.2 is therefore not a model in which the elements are functions, but a model that provides the coherence of a number of concepts.

8.5 Application of the Concepts

We shall now apply these concepts for clarity (see also Sect. 8.3).

The "Plague of Rats"

An old student flat suffers from a plague of rats. An investigation shows that there are 100 rats. A method is sought to kill these 100 rats, so:

$R_{intended}$ = 100 dead rats

Strategy:
Choose from the possible resources. Which resources are available?

- Bare hands
- Cannon
- Rat traps
- Poison

We must therefore examine the theoretical effectiveness (*R/R*) of each resource.

$\text{Effectiveness}_{\text{theoretical}} = \text{R}_{\text{expected}} / \text{R}_{\text{intended}}$

- Bare hands: $R_{expected}$ = 0 dead rats

$\text{Effectiveness}_{theoretical}$ = 0 dead rats/100 dead rats = 0

- Cannon, trap and poison. $R_{expected}$ = 100 dead rats

$\text{Effectiveness}_{theoretical}$ = 100 dead rats/100 dead rats = 1

Conclusion. "Bare hands" is eliminated as a resource because we can obviously not achieve the goal. Only the last three resources remain important. Which of these should we choose? The theoretical effectiveness of all three = 1.
This is why we must now examine the *theoretical productivity* (*R/S*).

$\text{Productivity}_{\text{theoretical}} = \text{R}_{\text{intended}} / \text{S}_{\text{expected}}$

Details of the cannon:

- We can borrow 1 cannon
- 4 people are required to operate it
- It takes 10 shots in 25 minutes
- Gunpowder + bullet = €2.50 per shot
- According to information provided by the owner of the cannon, 5 out of 10 shots usually hit the target
- 1 man-hour costs €12, so one man-minute costs €0.20

100 dead rats require:

(10 shots / 5 on target) × 100 = 200 shots

The sacrifice (the cost) for 200 shots is:

(200 shots / 10 shots) × 25 min × 4 persons × (€0.20/ man-minute) = €400

In addition, the costs of the gunpowder and the bullets are:

200 × €2.50 = € 500
Total $S_{expected}$ = €400 + €500 = €900
Productivity$_{theoretical}$ = 100 / €900 = 1 dead rat / €9 (or 0.111 dead rats per €).

Details of the trap:

- Purchase of one trap, €15
- 20 g of cheese bait per turn = €0.25
- Setting-up time per turn is 2 man-minutes
- Emptying time per turn is 3 man-minutes
- Costs are again €0.20 per man-minute.

For 100 dead rats we need to fill, set, etc., 100 times.
$S_{expected}$ = €15 + 100 × [€0.25 cheese + (2 + 3) man-minutes × €0.20] = €140
Productivity$_{theoretical}$ = 100 / €140 = 1 dead rat / €1.40 (or 0.714 dead rats per €).

Details of the poison:

- €1 per 100 g
- 1 g is enough for 1 rat
- They eat only 50% of the bait
- Placing the bait takes 2 minutes at €0.20 per man-minute

$S_{expected}$ = 100 dead rats × (100 / 50% × 1 g poison × €1 / 100 g) +
2 man-minutes × €0.20 = €2.40

Productivity$_{theoretical}$ = 100 / €2.40 = 1 dead rat / €0.024
(that is, 41.66 dead rats per €).

Conclusion. When the intended result is to kill 100 rats, *poison* is thus the best choice for the means, based on the theoretical productivity.

This conclusion appears logical, but is this the case? Haven't we forgotten something? We have indeed. We haven't considered the time period within which the goal must be achieved. Is that wise? Is the goal achieved in the same time by all three resources?

This time period must certainly be taken into account considering the speed at which the rats will multiply. This means reformulating the goal is required, such that it now becomes: *to kill 100 rats in one night* (lasting 10 hours). Does this influence the strategic selection of resources?

- *Cannon*: the same details as before.

  The number of shots that can be fired in these 10 hours is:

  10 hrs × 60 mins / 2.5 mins per shot = 240.

  According to the assignment, half of these shots hit the target. That is to say, 120 shots hit the target:

  $\text{Effectiveness}_{theoretical}$ = 120 dead rats / 100 dead rats = 1.20.

- *Trap*: the same details as before.

  With just the one trap only one rat can be caught each night.

  $\text{Effectiveness}_{theoretical}$ = 1 dead rat/100 dead rats = 0.01.

- *Poison*: the same details as before.

  It appears that when only one portion of poisoned bait is laid down, only four rats eat from it.

  The others do not eat from this bait when they see the effect it has. (8 g of poison is therefore enough; they only eat 50% anyway).

  $\text{Effectiveness}_{theoretical}$ = 4 dead rats / 100 dead rats = 0.04.

  Conclusion. The theoretical effect appears to be strongly dependent on the way in which we define the intended results. We conclude here that we must choose the cannon anyway. However, have we really exhaustively examined all of the resources?

  When we purchase 100 traps we can achieve our goal of 100 rats per night.

  $\text{Effectiveness}_{theoretical}$ = 100 dead rats / 100 dead rats = 1.

  If we do not use one portion of poison as bait but 25 portions of 8 g instead, then we can also achieve our intended result. As they only eat 50% of what is put out, we can thus expect 4 dead rats per portion of 8 g. So:

  4 rats × 25 portions of poison = 100 rats.

  $\text{Effectiveness}_{theoretical}$ = 100 dead rats / 100 dead rats = 1.

  We can achieve a higher result within the given 10 hours with the cannon (120 dead rats), but we do not need to do this. We must again make a calculation of the theoretical productivity for the sake of the strategic choice.

- *One cannon:* the same details as before.

  Required number of working-hours:

  100 dead rats / 1 dead rat per 5 mins = 500 minutes = 8.33 working-hours (instead of the 10 working-hours available).

For $S_{expected}$ for 200 shots, the same calculations are valid:
Total $S_{expected}$ = €900.

$\text{Productivity}_{theoretical}$ = 1 dead rat / €9 (= 0.111 dead rats per €).

Remark. It is assumed here that a cannon can be borrowed. This will probably not be the case in practice. The choice is then between purchasing or hiring. These costs should be calculated and included in the sacrifice. Buying a cannon that is not used afterwards means that the total purchase costs must be assigned in one go to this one project. Rental would then appear to be the cheaper option, since it gives the same result for a much smaller sacrifice. The productivity of that smaller sacrifice is therefore much greater.

- *One hundred traps*: the same details as before.

  $S_{expected}$ = 100 €15 + 100 × [€0.25 cheese + (2+3) minutes × €0.20] = €1625.

  $\text{Productivity}_{theoretical}$ = 100 / €1625 = 1 dead rat / €16.25 (= 0.62 dead rats per €).

  Intermediate conclusion. The cannon is clearly preferable over the 100 traps.

- *Twenty-five portions of 8 g of poison:* the same details and 2 man-minutes required to lay the portions of poison.

  $S_{expected}$ = 25 proportions × 8 g × € 1 / 100 g + 25 mins × €0.20 = €12.
  $\text{Productivity}_{theoretical}$ = 100 / €12 = 1 dead rat / €0.12 (= 8.33 dead rats per €)
  = $P_{standard}$

  Conclusion. Now the cannon appears to be more suitable for achieving the new intended result than the 100 traps. However, the strategic choice of the resource rests once again on the poison.

  Remark. We can rightly ask ourselves whether using 25 portions of 8 g of poison instead of one portion of 200 g are different resources in the strategic sense of the matter, or whether this is already a question of the tactics regarding the use of that *resource*, the poison. In this case we clearly find ourselves in an overlap area between strategy and tactics. We are, however, still busy making a choice based on the theoretical figures, i.e. with a strategic choice process.

  In practice, it is decided that the chosen resource (25 portions of poison) will be actually applied. The following morning, the worker reports that:

  $\text{Productivity}_{actual}$ = 4 dead rats per €.

According to our theoretical calculation it was Prod._{theor} = 8.33 dead rats per €, and this is set as standard for the practical implementation. There is obviously a big difference between theory and practice here. What has now actually taken place?

Is the intended result actually achieved, but for a much higher sacrifice? Or is it that the sacrifice is correct but the result is not achieved?

We therefore request the actual effectiveness and the efficiency (*S/S*). The answer is:

Effectiveness$_{actual}$ = 50%
Efficiency = 96%.

It can be concluded from this that the result has absolutely not been achieved, and in addition the actual sacrifice is greater than the expected sacrifice. These two figures together provide great insight, while the P_{actual} doesn't say much.

50 dead rats are actually found (some portions kill 4 rats, others 3, 2 or 1, and some kill none), so:

Effectiveness$_{actual}$ = 50 dead rats / 100 dead rats = 50%.

The poison was actually found to cost not €1 but €1.25 per 100 g, so:

S_{actual} = 25 portions × 8 g × €1.25 / 100 g + 25 × 2 man-minutes × €0.20
= €12.50.

$S_{expected}$ was €12. So:

Efficiency = $S_{expected}$ / S_{actual} = €12 / €12.50 = 0.96 = 96%.

In a one-off situation, post-assessment is useless. When this expected result occurs more often, than we need to see whether the actual effectiveness and efficiency cannot be increased.

Tactic. We must then see whether the tactic, the method of application of the resource, is correct (in the case of the cannon, it could be kept in a permanent position or moved around—i.e. hunting).

The worker has placed the portions of poison randomly in the rooms. From literature research it appears that the chance that the rats will discover the poison is far greater if the portions are placed at the entrances to the nests. From the literature it also appears that the attractiveness of a portion of poison is greatly enhanced by placing it on a piece of cheese.

From a methods study and better process planning, it appears that by choosing a more advantageous route, not 25 × 2 man-minutes = 50 man-minutes but just 40 man-minutes are required to place the portions of poison.

In addition, it appears that another shop can supply the poison for €1.10 incl. VAT per 200 g (the theoretical expectation was €1; the actual was €1.25).

Indeed, it appears from the following practical situation that the intended result (100 dead rats in one night) was fully achieved.

The total actual sacrifice = 40 man-minutes × €0.20 + €2.20 poison = €10.20.

Effectiveness$_{actual}$ = 100 dead rats / 100 dead rats = 100%.

Efficiency = €12 / €10.20 = 118%.

P_{actual} = 100 / € 10.20 = 1 dead rat / €0.102 (= 9.8 dead rats per €).

$P_{standard}$ = 8.33 dead rats per € (see before).

To control:

$P_{actual} = P_{standard} \times \text{effectiveness}_{actual} \times \text{efficiency} = 8.33 \times 100\% \times 118\%$
$= 9.8$ dead rats per €. This is therefore correct!

Remark. The costs of the cheese are left out of the picture here. Should you so wish, you can calculate the influence of this yourself.

In this example, the waiting time between laying the poison and the result does not increase costs. In practice, waiting time usually costs money. Also, these costs must then be calculated as sacrifice.

By improving the application of the chosen resources at the tactical level, both the effectiveness and the efficiency have been increased.

Based on our discussion above, it is clear that the concepts "effectiveness", "productivity" and "efficiency" are strongly dependent on the content that we give to the different types of results and types of sacrifices. This can clearly be the cause of much confusion and mutual misunderstanding.

References

Fayol H (1984) General and industrial management. Irwin, Homewood, IL

Hennipman P (1962) Goals and criteria (Doeleinden en criteria); Theorie van de economische politiek (Theory of the economical politics). Stenfert Kroese, Leiden, p 48

in 't Veld J (1993) Organisatiestructuur en arbeidsplaats (Organisation structure and jobs). Stenfert Kroese, Leiden

Chapter 9
Model for the Innovation Process

Abstract. When discussing function control it emerged that the evaluating function can arrive at the conclusion that the environment no longer needs the function fulfilled by the organisation. The primary function has been superseded. However, the organisation will usually still want to survive. What then? In the steady-state model there is no single function that anticipates that and does something about it. The model only covers the functions necessary for the execution and control of the *existing* processes. These processes are equipped to realise the currently defined goals. Missing from the model are functions geared towards innovation, towards *breaking through* to new goals, to perpetuating the goal "to survive". Using a comparable thought process, based on thinking in terms of functions and systems, we must be able to arrive at *new* goals for a pre-existing organisation. This innovation process is not a "one-off" affair. Because the environment is continually changing we must continually arrive at new goals in order to ensure the survival of the organisation. To this end, the functions to be fulfilled in that innovation process must be recognised and assigned to permanent organs.

9.1 Setting Up the Model for the Innovation Process

9.1.1 Environmental Reconnaissance and Definition of Goals

When the organisation's environment no longer needs the function the organisation fulfils, the internal goal "to survive" comes to the forefront. Survival is however only possible when new external goals can be found. That is why the organisation needs a function that scans the environment and attempts to describe the situation in that environment in terms of needs. Here, too, a desire analysis of possible interested parties is useful, but we will also have to apply other reconnaissance methods such as technology assessment and scenarios.

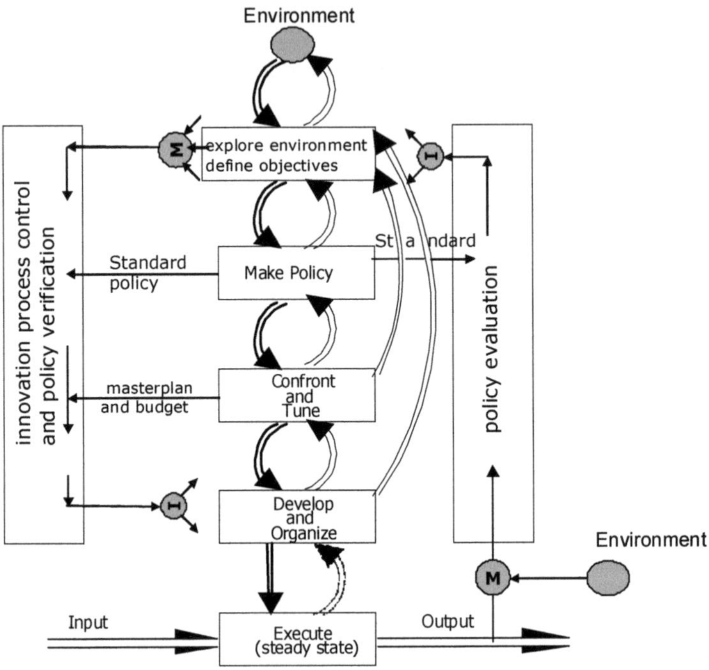

Fig. 9.1 The innovation model. Function model for innovation in an aspect system

From the needs identified, we can then formulate *possible goals* for the organisation. Considering the knowledge, people and means present, it is useless to formulate *all* of society's needs. Furthermore, we must remain within the domain of the existing organisation's possibilities. This will be returned to in the next section, on policy-making. We will denote these functions *environmental reconnaissance* and *definition of goals*. However, at this stage of the process, the definition of goals is not yet definitely possible. We must first weigh up other possibilities. The innovation process is not as straightforward as the steady-state process essentially is.

In this context it is useful to make a distinction between growth processes and stacking processes. This distinction originates from the analysis of "thinking as a process" by van der Goot and Malotaux. With *stacking processes*, the type, the time period and the sequence of the activities to be performed are known in advance. A *growth process* is however iterative throughout the whole structure. We are repeatedly faced with the task of developing new alternatives, choosing from these and testing that choice. It is crucial to a growth process that the project grows through each choice, but we must also solve the problems that are caused by these choices. Partial problems must be isolated and solved separately. Stagnation occurs regularly in this iterative process. We must return to a previous step because it appears that we cannot solve the problems caused by the last but one choice, etc. Maturation periods are continuously required to proceed, and this is also the case in the innovation process. We can compare this with a pilgrimage: three steps forward, two steps back, three steps forward, etc.

Ultimately, a combination of both types of processes is possible if a part growth process is also required within a stacking process. Conversely, part-stacking processes appear in each growth process. However, the innovation process largely has the character of a growth process and is therefore more difficult to put into model form than the "steady-state" with mostly stacking processes in the execution. The flow in the innovation model must often return again to a previous function in order to get back onto the primary process route. This is suggested in Fig. 9.1 by curved arrows, as opposed to the straight arrows used in the steady state model.

9.1.2 Policy-Making

After we have found potential new goals we must check the ways and means by which these goals can be realised. We must also set priorities for those goals. For various reasons we will not accept some possible goals, as true goals. Particularly in an innovation process in an existing organisation, it is important to take into account existing possibilities and not to take risks that are too great. To achieve diversification—i.e. adding other types of products to those that are currently being made—three important factors must be present:

- The development function should have specific knowledge and experience
- Production should have certain means as well as specific technological knowledge and experience
- Marketing (sales) should be focused on and have experience of the market for certain types of products for certain categories of customers.

We must, if at all possible, make a relatively gradual transfer to another type of product. Experience has shown that at each diversification step it is wise to ensure that there is still a relationship to the present situation for two of these three factors.

The 1980s saw an increasingly careful approach to diversification. It is now suggested that strategic decision-making should concern itself with:

- The range of the organisation's activities
- "Matching" the activities with the environment
- "Matching" the activities with the people and means in the organisation
- Allocating and redistributing the important people and means in the organisation
- The values, expectations and goals of those who influence the strategy
- The direction in which the organisation will move in the long term
- The effects of change throughout the entire organisation.

In short, it is about a good "match" between *opportunities and threats*. All of this fits well with the innovation model to be developed here.

In the different analyses we continue to use matrices in which we plot our own organisation and our own products as well as those of our competitors (benchmarking). From this, we determine our own strengths and weaknesses and we attempt to determine a strategy. Examples of such comparisons are:

- Market share against rate of market growth
- Competitive position against that company's attractiveness
- Competitive position against the phase of product/market development
- The maturity of the industry against competitive position (Porter 1986; Johnson and Scholes 1997; Gold 1991).

In *policy-making*, we examine which ways and means could be employed to realise those possible new goals. This process can again raise new questions for the environmental reconnaissance function. The returning arrow in Fig. 9.1 indicates this. Here too we encounter the question: how can we know whether all possible means have indeed been considered? To achieve this, the key is again to think in terms of basic solutions.

However, we cannot make a final new goal definition nor set a definitive policy until the discussion of the confrontation between these desires and the possibilities has taken place.

Remark. In Chap. 8 it was already suggested that we prefer to separate the specification of the goals from the determination of the policy. This primarily analytical separation is crucial to making sure that all possible organisational goals are discussed. Which goals eventually are or are not accepted is a result of discussion and/or balance of power between the parties involved. This is particularly true of conflicting goals. For competitive goals it is about which goals receive preference when the resources are limited. It is therefore important to make sure that no goal disappears out of sight prematurely. As soon as we begin to discuss the priorities between the goals and the ways and means by which we intend to achieve those goals, extensive differences emerge when interpreting and estimating the risk. For those reasons, in deference to an approach commonly seen in the literature, we have split the policy function in two: into a goals function and a policy function.

9.1.3 Confrontation and Tuning

We have now determined the ways and means that will be required for all sorts of possible goals in order to achieve such goals. We have stated our preference for certain goals. These are all still just *desires*. In order to be able to decide upon our definite choice of which goals with which policy, we must confront these desires with the organisation's *possibilities.* Does it possess the required knowledge, people and means? Can it develop these? And if the answer is "yes", within which time period; can the capital market be approached? And if "yes", within which timescale can the investments be paid back? Are organisational changes required?

To what extent can the changes take place simultaneously or will the different requirements have to take place in succession? Based on this *confrontation* of desires and possibilities, the prioritisation of the goals will perhaps have to be changed whilst other goals will fall outside of the possibilities. Various forms of information and new questions return to the previous policy-making function and the environmental reconnaissance function. This iterative process is usually followed a number of times before we arrive at a feasible tuning of the desires and the possibilities. Only then can a more definitive *goal definition* with a more definitive *policy choice* be undertaken. The most important goal of the tuning function is the *confrontation* between the desires and the possibilities for the appropriation of people and means. The final result of these series of iterations is an attainable *tuning* that we usually lay down in such documents as a *master plan* and *budgets*. However, this result is a derivative and not the primary function.

When finalising the determination of the goals and the policy, we should explicitly determine the demands on the system with respect to effectiveness, productivity, flexibility, throughput time, reliability of delivery, etc. In addition, the required degree and the manner of process control should be determined, as should the degree of centralisation or decentralisation, and the technologies, processes and organisational structure to be applied.

9.1.4 Development and (Re-)Equipping

The results of the previous process indicate the research and/or development that must be undertaken. This can have implications for:

- Product development: the development of products or services to satisfy the needs that have been identified
- Production development: the development of all that is required to both produce the chosen products or services in the most effective and productive manner possible and to deliver these to the environment
- Input development: the development of all required input factors, such as raw materials, people and means, capital, ideas and organisation
- Market development: the development of the output relationships, distribution channels and latent needs.

During this phase, we may not be able to find solutions for problems that have arisen, or we may find that provisional estimates of required time, capital and capacity were incorrect, that new questions arise or that unexpected new possibilities are discovered. This information and these questions can again lead to a review of the goals and policy, and thus to a return to previously chosen principle solutions. We must then make further new choices, after which a confrontation between desires and the (now better known) possibilities, will again takes place. In other words, a new iteration step is required.

Only when the possibilities are definitely known can the goals, the policy and the tuning be definitely realised. Then we can begin the actual equipping or re-equipping of new execution processes. This involves not only technological equipping and layout but also the eventual forming of new organic and personnel structures, the enlisting or changing of means and capital, and the training and eventual recruitment of personnel. The new or renewed executing process can then commence.

9.1.5 Control of the Innovation Process

Despite the typical iterative, growth process character of the innovation process, it will need to take place in a manner that is as controlled as possible. Thus, the setting, equipping and control of standards is also required here. In particular, it is necessary to "feel your way" through the first series of iterations between goal, policy and confrontation with the possibilities, and so these are difficult to plan. We will initially be able to make no more than procedural agreements. The network technique, which initially establishes relationships without estimating timescales, is best suited for this. In a more advanced arena, these relationship networks can be translated into network scheduling through the addition of estimated times. When the confrontation leads to an initial tuning of the desires and possibilities, the schedule can be made more concrete and can incorporate task-setting. The resulting (provisional) *master plan* also provides a schedule for continuing the innovation process. This innovation process is largely a one-off process despite the iteration steps. We can therefore only apply equipping, feed forward and eventually "add the (still) missing" as methods of control. A specific function for these controls must be incorporated. This function is symbolically depicted on the left hand side of Fig. 9.1. This black box for the function is not as detailed as it is in the steady-state model due to the complications associated with the iterative character. In Fig. 9.1, an information arrow is drawn to the left from the tuning function to the controlling function. After tuning is realised, the master plan and the budgets form the standards for the subsequent innovation process. This plan provides the programming of the timescale for policy implementation.

This master plan primarily specifies when actions should happen: the programming. This is best compared with the control of quantity in the steady-state model. However, it is then apparent by analogy that the control of the *quality* of the "product" in the innovation model is missing. In this case, however, the product is now a new policy and the implementation of it. Policy is formulated at the policy layer. In term of content, it looks similar to a specification of requirements. Then, as the innovation process proceeds, there must be continuous monitoring of whether the following steps adhere to that formulated policy, to that specification of requirements. Is this policy actually followed during the development and the equipping and the eventual implementation of the new process? Are emerging difficulties and impossibilities relayed to the previous functions, or is it necessary to improvise? Do personal gain and personal views play a role in the later decisions that occur at lower levels? We call this quality control the *policy verification*. This is also carried out by the controlling function. This is why, in Fig. 9.1, the arrow for "standards policy" is drawn from the policy-forming function to the left to the controlling function. This is therefore a quality standard that determines the policy of the subsequent content and the results of the innovation process. It may be tempting to view this as feedback, but this would be incorrect. It is actually feed forward, as we measure disturbances in the policy implementation and intervene based on the measurements, and so cause determines intervention.

Thus, only feed forward or add-the-missing are referred to by the control of the progress and by the control of the quality of the innovation process. Therefore, in Fig. 9.1, both the drawn measurements and the drawn interventions could occur at random somewhere during the innovation process (disturbance or deviation determines intervention). Both can concern the quality of the policy as well as the progress.

9.1.6 Policy Evaluation

At a certain moment in the innovation process, we make a definite choice of certain goals and create a specific policy. Whether that choice was the right one can only be judged later, when the implementation has been active for some time. We must also consider: are we really fulfilling an environmental need, or was that need purely fictitious or different to what we thought? Have we chosen the correct ways and means? In short, does this choice of goals ensure the survival of the organisation? Note that we are then back at the same situation discussed at the start of this chapter, where it was assumed that the existing goals were no longer valid. This is *policy evaluation*, and a function for this must also be incorporated into the model. This is drawn on the right hand side of Fig. 9.1. This is indeed feedback. In this way, the results of the innovation are measured and compared to that policy. However, when the results do not meet with expectations, then the interference entails a complete new cycle of the innovation process or at least part of it. This continuously renewed process of innovation is a metaprocess where feedback is indeed possible, in contrast to the one-off innovation process indicated in the model. Therefore, the metaprocess comprises a whole series of the models such as Fig. 9.1 in succession, and they sometimes overlap if other goals are also possible.

This metaprocess is essential for the survival of the organisation. Successful innovation is based on learning from mistakes. In this respect, Den Hertog (1988) introduced the concept of *learning ability*, which he defined as the ability of an organisation to collect relevant information, to select, to integrate, to store and to retrieve, to transfer and to use. To this list should be added: to interpret information. The metaprocess used in the innovation model is crucial to this learning ability.

The function *policy verification* in the model checks whether we correctly carrying out the innovation process, while the function *policy evaluation* checks whether we undertaken the correct innovation, and asks what can we learn from this.

9.1.7 Innovation and Improvement of Existing Processes

In the previous descriptions, the innovation process ended with the initiation of a new, equipped executing process. Problems can arise during execution: difficulties with the production, with the product, with the input or with the market. In the first place, it is the task of the functions in the steady-state model, as suggested in Sect. 4.3.1, to solve these problems. It is however possible that some development must occur in order to solve them. The upwardly curved dotted arrow in Fig. 9.1 indicates this. Sometimes this can also give rise to questions about new tuning or policy issues.

We once more perform a series of iterations through the innovation model. However, the difference in this case is that we are talking about a problem with a much smaller scale and range compared to the previously mentioned innovation process. In addition, the question has arisen from a pre-existing executing process, not from a reconnaissance of the environment for new needs. It is often difficult, in practice, to precisely define the boundary between actual innovation and large improvements in the existing situation.

The innovation model drawn in Fig. 9.1 can thus apply when starting an organisation from scratch as well as innovating an existing organisation and improving the execution of existing processes. This model is also valid for each aspect-system separately.

In principle, the model was built from a "zero situation"—there was no goal and we went in search of needs to be fulfilled in the environment. Based on that environmental reconnaissance, the model was further developed downwards.

This innovation model is, with various adjustments, based on Malotaux's "main functions" model. He developed this model for use in his organisation consultancy business. He ascertained that many companies found it very difficult to implement deliberate and controlled innovation. They often had sufficient command of the steady-state process but failed to implement new developments.

9.2 The Nature of the Model for Innovation Processes

A model is an aspect system of a higher aggregation layer used for studying another system. Depending on the goal set by the researcher, a system is a distinguishable collection of elements within the total reality. These elements have mutual relationships and (eventual) relationships with other elements in the total reality. These respective definitions given in Chap. 2 prompt the question of which relationships the steady-state model and the innovation model actually contain. Here, the elements are functions in processes that must be fulfilled in the system. For each of these functions a process is again necessary at a lower aggregation layer. In the innovation model the eventual new primary process is represented by a black box. This is thus the process by which the system must derive its right to exist in the future and by which the income for the system must be generated. All of the other functions in the models can, in turn, only derive their rights to exist from the contributions that they can make to improving or simplifying the functioning of the primary process at the operational level. The key to the connection between the functions is obtained in this way. The relationships between the functions in the model are therefore formed from the items that the processes that fulfil the functions provide to each other. The outputs from the functions are the connecting links. These outputs form the inputs for subsequent processes. A function is then also described in terms of its desired contributions, and not in terms of how that contribution is realised (and so not in terms of throughput). The models show functions in terms of their functional coherence; that is to say, in terms of the required effects on or their contributions to each other. This involves the mutual tuning of processes via the required contributions (outputs) in order to realise a common goal. The models do not describe tasks; they abstract the different ways in which a function can be fulfilled. Both the innovation model and the steady-state model are *empty* with respect to the contents of the processes. Indeed, very different contents of *concrete* processes can flow through these models. It is this property of abstraction that makes these models generally applicable. They are only normative for the coherence of the outputs from the different contributions. Using these models then we can explore totally different primary processes at our own discretion: the production of coffee grinders or aeroplanes; the healing of patients or the education of academics.

The models are also valid for one aggregation layer below each process that fulfils each function that arises in these models: this is the "nurse's effect". The models can therefore be applied as a construction module in the design of an organisation. However, it is important to delineate the different aggregation layers at which the models are applied. They are thus *conceptual* models that support theoretical development.

9.3 Policy Evaluation

In the innovation model, the organisational goals are determined at the goals layer. At the policy- and tuning layers, the ways and means to realise these goals are determined, broken down into steps, set out in time and confronted with the possibilities. This results in (among other things) a master plan and budgets; the as-it-should-be situations at the time; the standards that must be adhered to. The control loop for policy evaluation must compare the as-is-situation with those standards in the future. In doing so, they must not only look inside the organisation but must also be attuned to what is going on in the environment, eventually via other organs. No matter how well a certain organisation may seem to function when viewed in isolation, it may not function quite so well when viewed as part of a greater system. An organisation's performance should be measured in some manner. The performance is *not* the quantity of output; this is, at most, one aspect of it. It is about how well its function is fulfilled, its effectiveness and the sacrifices that must be made to fulfil its function (the productivity). We should take into consideration not only the quantity of output but also its quality, the adherence to delivery times, the degree of service, etc. The organisation must optimise various factors in combination, such as the effectiveness, productivity and efficiency of the product flow, as well as the quality, reliability, throughput time, etc., of the product. The criteria for these must be deduced from the pattern of norms and values that are active in the environment. For example, consider the current changing opinions on the required life expectancy of products, the importance of environmental hygiene, working conditions, type of work and working climate. These norms and values in the community are not objective; they are not absolute and unchangeable. They are based on a multitude of aspects: economic, social, ethical and political factors.

The concepts of effectiveness, productivity and efficiency are important when testing these standards. Therefore, we should not (as is often the case) only apply them in technological and economic aspectsystems. We must also use them in the social aspectsystem, in the information aspectsystem and in other aspectsystems.

The question is in fact more about whether the results fulfil the goals, and whether those results are achieved with the lowest possible sacrifices. When we make the strategic choice of ways and means in the policy, we continuously encounter the problem that we must first know whether something *can be done*. Only then can we test the *desirability*, after which we can consider whether we should do it. This last consideration is partially based on the sacrifices to be made. We then consider whether the desire for it is so strong that we consider that the sacrifices that must be made to do it are acceptable. First we test the desirability of doing something, based on a general idea of what should be desirable. Only then can we develop methods with which we can do it. After this, we determine the expected sacrifices so that we can then retest the desirability, taking into consideration these sacrifices to be made. Monetary values are often used for these comparisons, but money is not very useful as a measuring unit for inputs and outputs of aspectsystems other than economic aspectsystems—for example, inputs and outputs of human motivation, energy, work satisfaction, product reliability and reliability of production processes—and for the ramifications of these types of aspects on the future of the organisation. In this case we must still work with the concepts of effectiveness, productivity and efficiency, but these are then expressed in units other than money; units that are often not expressible quantitatively, and can only be ranked (for example, examination grades) or are only attributive (for example, yes/no or red/green/blue). During policy evaluation we need to test all of these factors and their relationships to standards and the needs of the environment. However, we cannot quantitatively compare many of the aspects involved; for example, money and work satisfaction, so the final comparison with the standards can never be performed based on one indicator—it requires a complete series of (often subjective) comparisons.

References

Gold B (1991) Technology in corporate strategy. Int J Technol Manag (Spec Publ)
den Hertog JF (1988) Technologie en organisatie; mythe en missie (Technology and organisation: myth and mission). Edition NKWO, The Netherlands
Johnson G, Scholes K (1997) Exploring corporate strategy. Prentice Hall, Englewood Cliffs, NJ
Malotaux PChA (1997) Hoofdfuncties in een onderneming (Main functions in an organisation). Lecture notes 111 in Business Management, TSM Business School, Enschede
Porter ME (1986) Competitive advantage. The Free Press/Macmillan, New York

Chapter 10
The Design Process with the Conceptual Models

Abstract. Where should the models be positioned in the design process? We start by examining the design process of an industrial system itself. It appears that the functional design should be done in an interdisciplinary way in order to construct "shared memory". In contrast to developments in design theory, every domain should be involved and should start from the same objectives. Both the PROPER model and the steady-state model support this interdisciplinary design approach. The emphasis when using these models will shift from "streamlining the existing" to "starting from scratch" in order to make use of new opportunities. The result of the process will be a function design that all domains can draw upon for their detail design. The function design can be completely expressed in terms of both function models and may provide a source of information for future designs.

10.1 Introduction

In Chap. 5, the process–performance model (PROPER) for an industrial system has been defined. This model represents the interrelated aspects (orders, products and resources) that arise during the design process, not only from an operational perspective but also from a control viewpoint. The use of the PROPER model during the design process will be examined in this chapter.

During the design process, the level of detail in the model gradually increases with succeeding cycles of problem formulation and solution. The elements of the model remain PROPER models at different aggregation layers during this trajectory. As such, the model is a frame of reference for all disciplines involved. The model plays an important role in problem formulation: each solution or each decision formulates new problems in more detail or tapered to one or more subsystems or aspects.

The design process itself will be examined first. The nature of design has changed significantly, not just as a result of a changing market, but also as a result of developments in information technology. The emphasis in the process has shifted from "streamlining the existing" to "starting from scratch" in order to make use of the new opportunities.

In this chapter, the application of the PROPER model to the design process will be explained via the innovation model (see Chap. 9). The elaboration of the PROPER model is then represented then as a process composed of two clearly distinguished steps: function design and process design. Each of these design steps consists of successive cycles where each cycle determines the context, function, structure and behaviour, in this order. When changing from function design to process design, the phrasing of the questions changes from "what?" and "why?" to "how?" This transition coincides with the transition from generic PROPER modelling to the specific modelling of each discipline. The transition will be illustrated for the fields of organisation, technology and information.

The PROPER model will always be required for communicating about and feedback on detailed designs. The use of the model supports the structured recording and evaluation of elaboration and changes. This satisfies the major aim of constructing a "shared memory" from which future design processes can then draw. Shared memory starts with "shared meaning", and the PROPER model plays a major part in this (Konda et al. 1992).

At the end of this chapter, the steps of the design process in which behaviour descriptions can contribute to decision-making will be explained. This provides an introduction to modelling time-dependent aspects of the industrial system.

10.2 The Design Process

The pressure to reduce costs and lead times, to upgrade quality and to enhance the effectiveness and productivity of industrial systems in general has increased enormously during the last few decades. As a result, the importance of (re)engineering has also grown (Davenport, 1993). Both the number and the kinds of tools available for design have also expanded due to the development of information technology. For example, computer-aided design (CAD) and computer simulation are commonplace now. Beyond this, information technology has also changed the content and structure of industrial systems. The increased speed of communication has influenced decision-making processes in industrial systems, and technological tools have become more advanced through the use of automation. Therefore, the design process is increasingly being from scratch, resulting in the risk of errors being repeated or "re-inventing the wheel". To rule out this risk, design decisions and principles should be recorded in terms of the PROPER model, since the model is conceptual and fundamental with respect to functions and processes. Therefore, this model must be positioned in the design process. To determine this position, the characteristics of the design process of an industrial system must be described using a model.

Several conceptual models for a design process have been developed. In a classification developed by Konda et al. (1992), it was concluded that these models are usually developed from the viewpoint of a single discipline. There are models for the design of technical systems or products, for information systems and for organisations. An industrial system however covers all of these aspects.

Currently, the design process is not restricted to the just the product. In all disciplines, the "making of" the product and its use are also included in the design process from the very start. The (efficiency and effectiveness of the) manufacturing process is considered during the design of products, and its utilisation during production (e.g. maintainability) is considered during the design of a technical system. The design process of information systems includes the implementation trajectory and maintenance, and the design of an organisation is not complete without including this process. Terms like "concurrent engineering" and "life-cycle engineering" all refer to this tendency to integrate different aspects of the product.

The common starting point for all design processes is usually the function to be fulfilled. After that the function is detailed and finally concretized for implementation.

In order to support an interdisciplinary approach to the design of an industrial system, a step is required before the different (parallel) design trajectories of each discipline. We consider the design of an industrial system to be a combination of the design of a product (the industrial system itself) and the design of a process (the way in which the industrial system works).

Product design involves determining the objectives of the system. Central questions in this process are "what is feasible?", "what is required?" and "what functions need to be fulfilled by the system?" It is a strategic decision process and it will be termed the *function design* from now on (see Sect. 10.3.).

Specifying the way in which the system will work involves defining structure, processes and resources, and this will be termed *process design* (see Sect. 10.4). Process design deals with the optimal utilisation of resources, technology and information and therefore belongs to the field of tactical decision-making.

Function design covers the performance part of the PROPER model, while process design covers the processing part.

Function design distinguishes between innovation and improvement. Improvement usually only concerns the reorganisation or reengineering of an existing system, which implies a rearrangement of existing functions or a different interpretation of these functions. Innovation concerns the extension, reduction or modification of functions due to the introduction of new technology, resources and/or organisation.

Jonas (1997) even distinguishes between three steps in the design process: analysis, projection and synthesis (see Fig. 10.1). Jonas states: "Transforming a vague feeling of discontent into a solution turns out to be a three-step process of reducing uncertainty (contingency). The traditional concept of industrial design neglects the first two steps and acts at the very end of the process". The function design as described above corresponds to the first two steps according to Jonas. After these steps, a "problem" is formulated that can be used for process design. During the first

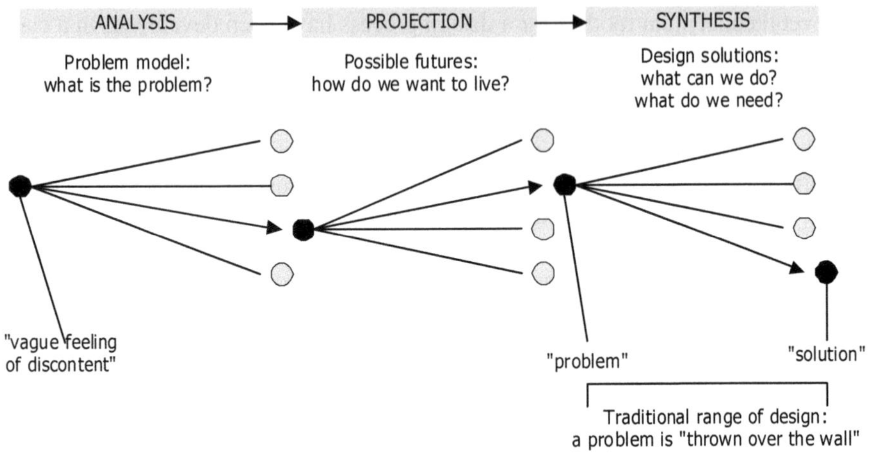

Fig. 10.1 Design as a three-step process of "problem solving"

step (analysis), a feeling of failure or discontent (e.g. "the market asks for something different") is translated into a number of possible reasons (e.g. "we are too expensive", "we don't make the right product", "we don't deliver in time").

In the step called "projection", the feasible and desired possibilities that could remove these reasons (such as "expand", "shrink", "innovate processes", "innovate products", "automate") are investigated. Deciding on these possibilities results in a definite problem formulation for the design trajectory.

Both the division of the design process into function design and process design and the division into analysis, projection and synthesis are plotted in relation to the functional steps of the innovation model (as described in Chap. 8) in Fig. 10.2. To make comparisons easier, the environment and the functions "verification" and "evaluation" have been omitted (see Fig. 10.2). On the right hand side of the figure, the result of the design process is represented: an operational industrial system. This system is represented by a PROPER model.

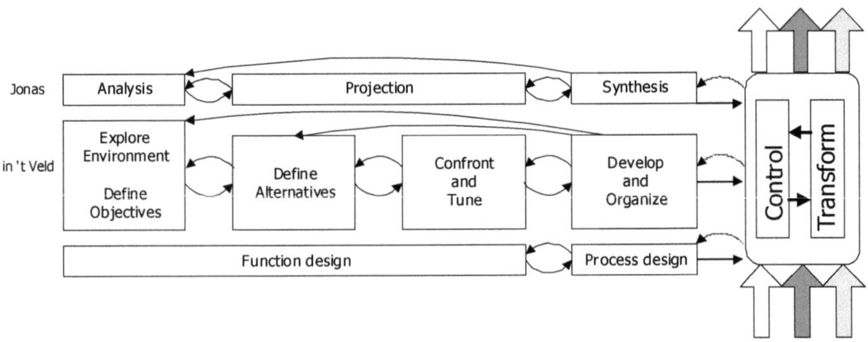

Fig. 10.2 The innovation model as a design process

The second step in the innovation model is called "define alternatives", in order to emphasise the iterative character of the process. Originally we used "make policy", but a policy (or a selected alternative) is achieved after several iterations via confrontation and tuning and (provisional) development.

The figure shows that function design encompasses the steps "explore environment and define objectives", "define alternatives" and finally "confront and tune". The process design takes place in the step "develop and organise".

Most of the literature on design focuses solely on process design. However, function design involves determining what can be required from the system, and this is the starting point for process design. This distinction correlates well with the different performance indicators described in Sect. 8.4. During function design, the strategic indicators of theoretical effectiveness (Eq. 8.1) and theoretical productivity (Eq. 8.2) should be determined. This cannot be achieved without iterating parts of the process design (albeit provisionally) because the expected efforts are greater than the resources allow. Having determined the intended results, the results that are expected and the efforts that are expected, a definite configuration can be selected, which will be fully elaborated during the process design.

Example 10.1. The Automated Container Terminal Project (see Chap. 11) was a design trajectory to develop a largely automated container terminal. This project was preceded by several years of research to determine both the resources and technology that were most appropriate and feasible for achieving the intended results. In those years of research, for example, several alternatives were investigated for the quay transportation system, such as rail-bound transportation, conveyors and free-ranging automated vehicles. In the end, a system with automated vehicles was selected based on the requirements of flexibility, productivity and technical feasibility. How to construct these vehicles (with respect to automation and technology), how to control them, and even which part of the functionality required on the sea side they needed to fulfil had not yet been determined. It was clear, however, that this system could offer the best theoretical effectiveness and theoretical productivity.

10.3 Function Design

Function design aims to determine a system configuration that is able to provide a required performance in an optimal way; i.e. to provide the intended result with acceptable efforts. For an industrial system, this can generally be expressed by: "offer the required service with acceptable costs" (Lambert 1997). During function design, the system configuration should be tested for feasibility and desirability.

The steps in function design will be denoted by the functional terms of the innovation model shown in Fig. 10.2.

Step 1. Explore the Environment and Determine the Objective

During function design the environment mainly consists of the customer market, the society, the company to which the system belongs and the chain to which the system belongs. By exploring the environment, the need of the environment (the required service) is determined.

This need is translated into objectives, preconditions and principles. Preconditions fall outside the range of influence of the system and are firm restrictions on the rest of the design trajectory. Examples are statutory regulations, environmental laws and the like. Principles are conditions drawn up by the company's culture. They can be influenced, but changing a principle goes beyond the scope of the system to be designed and often involves high costs.

Objectives are expressed in terms of the PROPER model:

- Regarding the order flow: what is the demand composed of, what is the required lead time and what is the required reliability of delivery?
- Regarding the product flow: what are the products required, what is the required quantity and quality and what are the costs and flow times?
- Regarding the resource flow: what is the required quality and quantity of resources with respect to the order and product flows?

In terms of the relationships between the flows, this leads to questions such as:

- Between the order and product flows: what is the ratio of the order quantity to the product quantity?
- Between the resource and product flows: what is the required flexibility of utilisation?

The objectives for the order and product flows strongly influence the resource flow. In cases that involve, for example, large seasonal influences (e.g. the sugar industry), continuous operation (e.g. service departments) and dangerous work (e.g. the chemical industry), social factors play a major role. In addition, the increasing degree of automation has enabled new modes of operation and working methods, but these have not always been welcomed and they have led to considerable resistance to innovations.

At the end of this step the objectives are expressed in terms of the intended results.

Step 2. Define Alternatives

During this step, a number of alternative configurations are determined that may satisfy the requirements. Defining alternative configurations involves thinking creatively and structuring. Creativity does result in new ideas, but the results of it can and should be reflected in terms of the following structured approach.

Defining alternatives is actually an iterative process that can be defined globally as a repetition of cycles consisting of context determination, function determination, structure determination and behaviour determination (Ackoff 1971).

Table 10.1 The cycle of determining alternatives

Context determination	In terms of the systems approach, this is the determination of the system boundary. The system is considered part of a larger chain. Upstream, the system can be expanded in the direction of suppliers, downstream in the direction of customers. It is also possible to shrink the system. *System concepts* will result, in which different configurations can be selected. Each alternative has a primary function, based on the objectives.
Function determination	The primary function is divided into subfunctions (zooming). The intended result is defined for each subfunction.
Structure determination	Subfunctions can be particularized in the horizontal and vertical directions. There are two major ways of doing this horizontally: • Specialization/parallelization: the flows are split or combined • Differentiation/integration: the functions are split or combined. This step is a vitally important one, because splitting and combining flows and functions influences both the results and the efforts that can be expected. Splitting functions or flows generally results in increased costs (extra transfer actions, increased space requirements and the like). Combining flows may also lead to increased costs (complex technology, turnaround costs, extra sorting, etc.). The particularization in the vertical direction concerns the control structure. Control echelons are introduced and the degree of autonomy for each function group is determined. This type of structuring indicates the controllability and control burden (the need for control). A number of *structure concepts* are defined as a result. For each structure, the results that can be expected should be determined.
Behaviour determination	Each structure causes "behaviour". This presents itself on the one hand in communication and consultation requirements, and in the other in time-dependent phenomena with respect to the contents of a structure (stocks, throughput times, etc.). This step therefore determines the expected behaviour: a *behaviour concept*. Behaviour also influences both the results and the efforts that are expected. Each structure exhibits its own specific behaviour.

Each successive cycle takes place at the next aggregation layer. The following table shows the contents of each part of a cycle.

Table 10.1 shows that a feeling of discontent is translated into configurations consisting of a system and structure concept with a corresponding behaviour concept. It leads to a problem formulation for process design. In terms of Jonas's model, the analysis yields system concepts and the projection phase yields structure and behaviour concepts.

The PROPER model is the basis for defining system and structure concepts.

During the step of defining alternatives, each of the disciplines involved should evaluate the feasibility of a configuration from its own perspective. The model is not bound to a specific discipline and clearly reflects the environment, functions and structures. It is considered a "cognitive map" of the design. According to Dwyer (1998), such a map is "a system model from the perspective of how people involved with it will understand it". He further states that "systems incorporating human beings must be designed with the cognitive properties of humans in mind".

The determination of results to be expected is actually a part of process design, albeit a provisional one. Through draughts, prototyping, experience and simulation, each discipline contributes to this determination. This illustrates the iterative character of design.

If a system or structure concept is considered unfeasible (the results that are expected don't match up with the results that can be expected), this concept will not be elaborated any further. For the remaining configurations, the intended results and the results that are expected are now specified to the layer of subfunctions.

The PROPER model does not reflect the behaviour concept. It is a static model of the system. Step 3 will address this further.

Step 3. Confront and Tune

With the intended results and the results that are expected in hand, this step aims to determine the efforts that are expected for each configuration. To do this, the process component of the PROPER model is examined. A process takes time and capacity (costs). Confrontation means, for each discipline involved (in this book these are technology, organisation and information), separately assessing "can we do this?" and "do we want this?" All of the disciplines together then try to adjust to one another. For illustration purposes, common questions to be answered for each alternative are formulated for each of the disciplines in Table 10.2.

The results of "confront and tune" include a specification of the results and efforts to be expected for each alternative. The alternative with the maximum theoretical productivity is selected first. The other alternatives are retained in case it is found that this alternative does not match expectations during process design.

To determine the theoretical effectiveness and productivity, an understanding of the behaviour of the system is required. Simulation is an outstanding tool for achieving this.

Even without concrete resources being defined, simulation enables the expected behaviour of the processes in the PROPER model to be evaluated. The behaviour of functions can also be represented. In Chap. 5, the notion of "behaviour" was defined exactly.

Table 10.2 Confront and tune by discipline

Technology	Are the grouped functions technologically feasible? What kinds of hardware (and developing time and capacity) are required? What are the consequences with respect to operations, maintenance and environment?
Organization	What will the departments be? Are we able to and do we want to realise these within the existing organisation? What are the demands on the competencies of people and other means? Can they be obtained here or elsewhere? What educational efforts are required?
Information	What are the demands on architecture, software and hardware? Which administrative systems, control systems and production support systems are required? Are we able to develop the systems required in-house, or should we hire capacity for this?

10.4 Process Design

10.4.1 Introduction

Process design starts at the point where a configuration is selected. The selection may be provisional, as it is when iterating the "confront and tune" function, where different configurations are compared. The system must now be developed and organised. In function design, the function structure is reflected upon, including the intended results and (a first estimate of) results and efforts that can be expected. These values are the target figures for the optimal process design.

Jonas (1997) notes that this is the territory of the traditional design approach. Process design is a multidisciplinary/monodisciplinary approach rather than an interdisciplinary trajectory. The methodologies of all disciplines currently contain a stage called "conceptual design", probably based on the multidisciplinary requirements of the environment to which the methodology is applied. This conceptual design shows a large overlap with the function design of the preceding paragraph. The functions that are defined, however, usually cover just one aspect or subsystem of the PROPER model and use terminology that originates from the discipline itself. If we represent it by the innovation model structure of Fig. 10.2, the multidisciplinary approach results in Fig. 10.3.

The interdisciplinary approach used so far here takes all aspects and a consciously chosen system boundary into account during function design. The result is shown in Fig. 10.4.

The characteristic design steps will be described briefly for each of the disciplines of technology, organisation and information, and connections between function design and process design will be illustrated in the following sections.

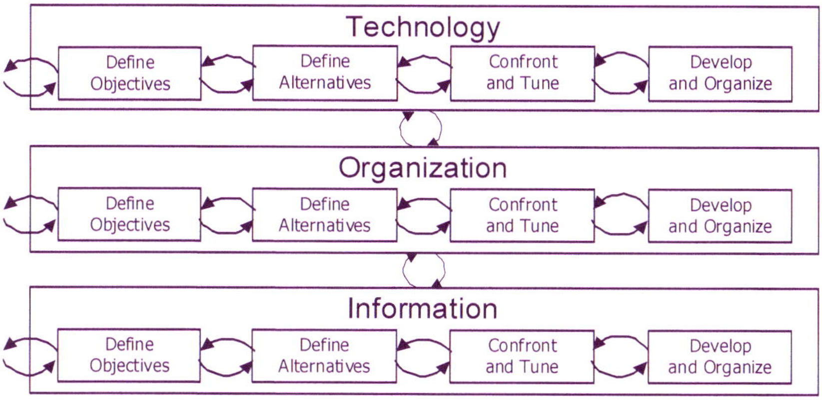

Fig. 10.3 A multidisciplinary design approach

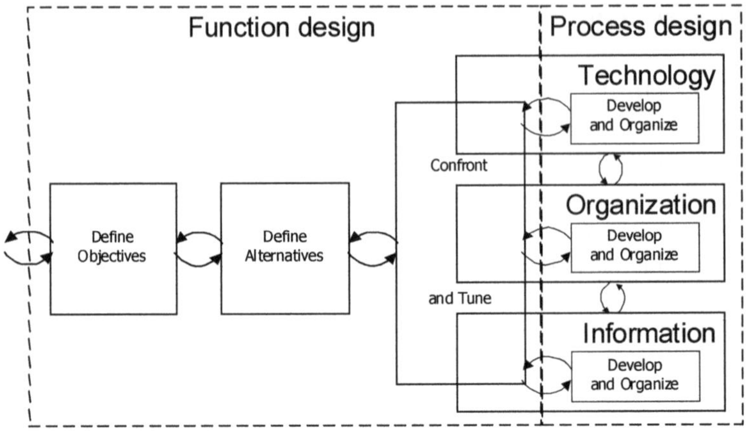

Fig. 10.4 The interdisciplinary design approach

10.4.2 The Design of Technical Systems

Technological process design mainly focuses on the resource flow of the PROPER model. It concerns the design of machines, tools and transportation equipment. Function groupings that were established during function design should be physically filled in at this stage. Doepker (2001) distinguishes the following steps:

1. *Establishing needs and generating ideas*, resulting in requirement specifications (the functional and design requirements and design criteria). Functional and design requirements have already been formulated with the PROPER model. The design criteria are described by Pugh (1990), and they consist of the product design specification (PDS), the organisation and the personnel. The PROPER model is an appropriate way to reflect these criteria.
2. *Conceptual design* is the decision-making performed to obtain a concept. Doepker says that this is a very important step because design changes are still "cheap" to implement (note that he is referring to changes in a single design here).
3. *Preliminary or detailed design*. Pahl and Beitz (1996) call this "embodiment design"; layout and form are quantified during this step. Materials and geometry are defined.
4. *Final design*. Detailed analyses (stress analysis, shaft design, heat transfer, etc.) are performed during this step.
5. *Implementation*.

During steps 1 and 2, specific requirements are added by the discipline of technology in relation to materials, safety, size, weight, ergonomics, etc. Steps 3 and 4 detail the design, eventually resulting in extra restrictions. These restrictions are fed back to the other disciplines since the functionality in the PROPER model is defined collaboratively. More detailed insight into "behaviour" is added to the model.

Example 10.2 illustrates the combination of the product flow and the resource flow from the PROPER model in order to achieve a complete functional specification of the requirements for the technical resources. These form the basis for confrontation and tuning and are the starting point for technical design.

Example 10.2. The process of importing containers at a deep-sea terminal includes the process of transferring the containers between the ships and the stack area. It is decides that the resources to be used will include quay cranes to unload, automated vehicles for transport and stacking cranes to stack the containers. Transfer functions are required to transfer containers between the resources.

Suppose that the product flow for this part of the Operate function is modelled as in Fig. 10.5 from (Veeke and Ottjes 1999).

One of the alternatives that needs to be investigated during the confront and tune step of function design is the one where the automated vehicles don't have a lifting installation; the transfer functions are to be performed by the other equipment.

Adding the resource flow to the product flow results in Fig. 10.6.

Now the technologists are asked to provisionally work out the vehicle design. The required functions are defined by isolating the vehicle part from Fig. 10.6 and zoom into the assignment part. Note that not all of the control functions are shown in the figure to make it easier to follow.

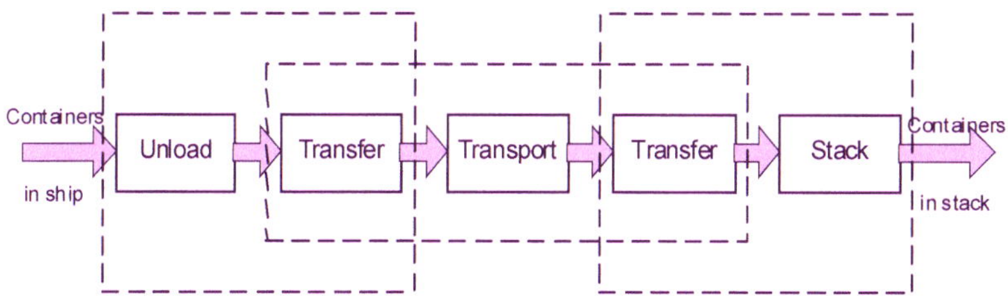

Fig. 10.5 Functions in the container import process

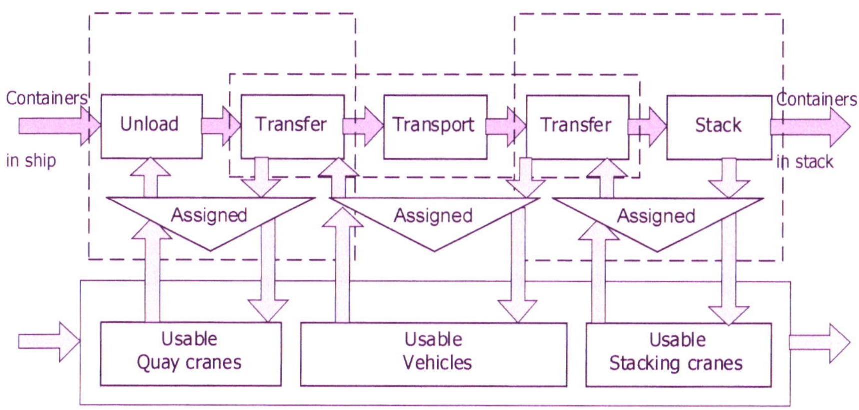

Fig. 10.6 Product and resource flows in the container import process

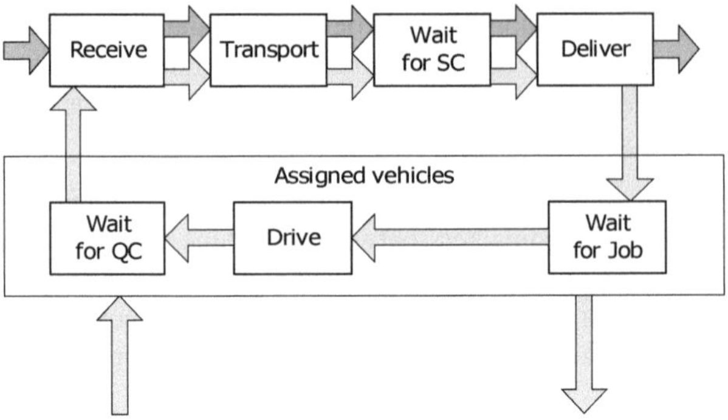

Fig. 10.7 Vehicle functions in the container import process (QC = quay crane, SC = stacking crane)

From the technological viewpoint, the size, speed, etc. of the vehicles are major design criteria. Vehicles always have a physical position, use space and they should be "stored" when they are waiting for a job or for transfers to be performed by quay crane and stacking crane. This becomes clear if we zoom into the buffer for assigned vehicles (Fig. 10.7).

The Receive, Wait for SC and Deliver functions are parts of the transfer functions shown in Fig. 10.6. The results and efforts to be expected with this type of vehicle in this configuration should be estimated from these functions. Expected results will be the number of transports per time unit, while efforts will be the number of vehicles, their occupation and space requirements.

The Wait functions can be fed back into the overall design to allow for comparisons with other alternatives.

The purely technological functions Drive and Transport will be worked out after we have chosen between the alternatives. In this case, the results and efforts to be expected are accepted and become standards for the technological design.

10.4.3 The Design of Organisation Systems

Bikker (1995) defines a strategy for organisation design (Fig. 10.8). He distinguishes between two major design pathways:

1. The analysis of objectives and policy
2. The analysis of the organisation of processes.

The first pathway—analysis of objectives and policy—coincides fully with the function design of the design process in this book. The PROPER model is very useful for investigating the existing situation and designing a new situation, which is the second path. Beyond this, the model is able to visualise policy decision problems in a compact and conveniently arranged way. Example 10.3 illustrates this.

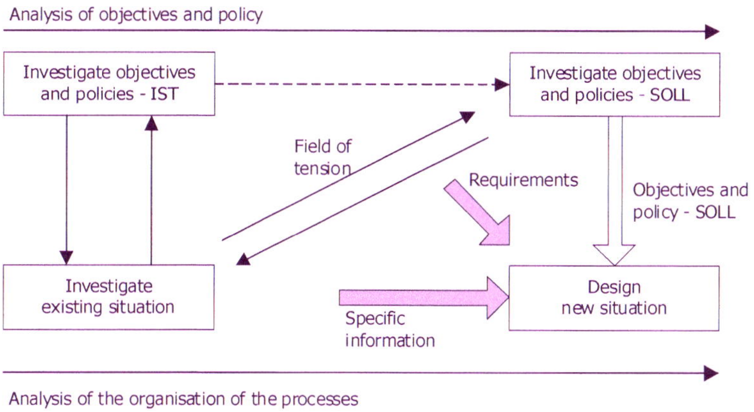

Fig. 10.8 Strategy of organisation design

Example 10.3. The Carrier refrigeration company considered carrying out a major reorganisation in order to facilitate the combination of different sales companies operating in a partly overlapping market. The characters (cultures) of the companies varied from purely sales-oriented to heavily installation-oriented. Products ranged from simple refrigeration equipment to complex installations. The installations were maintained by the company during their lifetimes based on a service contract with the customer.

A PROPER model for this company was created at the highest level of abstraction, such that the order and product flows were visualised (Fig. 10.9).

Four main functions can be distinguished: a sales function to sell new installations and maintain customer relations; a service function to respond in time to disturbances; an installation function to perform the installation process; and a maintenance function to maintain installations.

The model shown in Fig. 10.10 was used to illustrate and discuss two principal organisational alternatives.

Each alternative has consequences for the rest of the design process, the personnel and the flexibility with respect to in- and outsourcing. The interfaces between obtain/secure and commerce/engineering were particularly big discussion points.

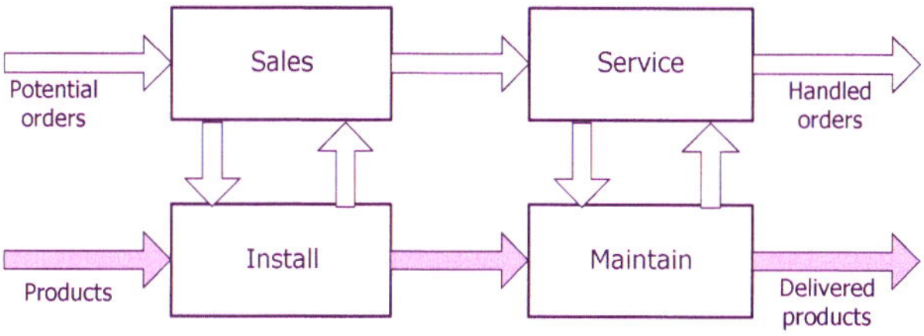

Fig. 10.9 Main functions of the Carrier refrigeration company

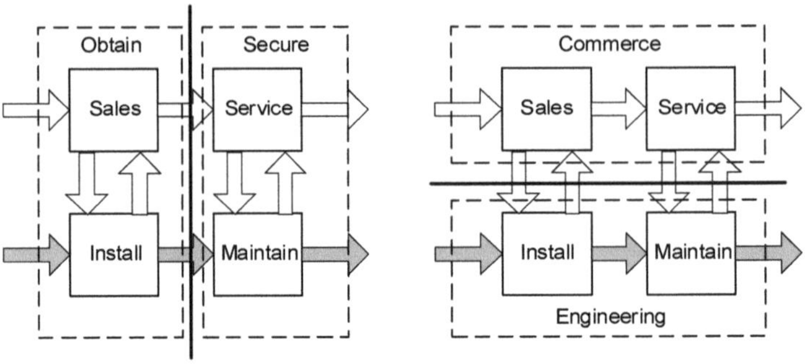

Fig. 10.10 Alternatives for the organisational structure

The PROPER model represents a conceptual organisation design as a starting point for process design. In process design the consequences of function particularisation are investigated, initially to compare different alternatives and then later to detail the selected alternative. The departments should be decided upon (the organ structure), as well as the competencies and the responsibilities (the personnel structure). These all determine the resources required.

10.4.4 The Design of Information Systems

Burback (1998) distinguishes between four fundamental phases that occur in all software engineering methodologies: analysis, design, implementation and testing. These phases address what needs to be built, how this will be built, actually building it and ensuring that it is built to a high quality. These apply to not only product delivery, but also to deployment, operations, management, legacy and the discontinuation of an information system.

The analysis phase defines the requirements of the system, independent of the way in which these requirements will be accomplished. The PROPER model can be used for this purpose. If one replaces the "order, product and resource" flows by "information on order, product and resource" flows, the model completely reflects data flows. In addition, the use of the PROPER model prevents the use of different element types during requirement analysis. "Documents" will be assigned to data flows (instead of functions); databases will usually be positioned at stock/buffer locations and within evaluation functions. One should however be aware of the fact that the PROPER model is only concerned with operational data. Data used for innovation and modelling purposes are not contained in this single design approach.

The next design phase concerns the structuring of the information system: the architecture. It defines components, their interfaces and behaviours. They can be derived directly from the PROPER structure. In the implementation phase, components are built either from scratch or by composition. Finally, the testing phase ensures the correct quality level of the system.

Most methodologies currently provide object-oriented development during the analysis and design phase. According to Lodewijks (1991), object orientation increases the level of abstraction of programming languages and enables a better fit of the design process to the human perception. He shows a direct projection of a functional design of a physical process or system to object-oriented programming (Lodewijks 1996).

10.5 Simulation as a Supporting Tool for the Design of Industrial Systems

In the last few sections, simulation has been mentioned several times as a supporting tool for decision-making during the design process. In both function design and process design, simulation can be applied in conjunction with the PROPER model to "quantify the structure". A logistic system is a complex system consisting of a large number of elements with multiple interrelations and often stochastic behaviour. In these situations, a static structure (such as the PROPER model) is not sufficient for deciding on the most appropriate design of a system. Dimensions should also be taken into account and control algorithms must be tested thoroughly. The complexity of the system makes a mathematical description impossible. Simulation is however capable of describing the time-dependent behaviour of these systems and performing experiments as if the system is (virtually) operational. Simulation can even provide enhanced insight into seemingly simple situations, especially for stochastic phenomena. Global models used during strategic decision-making are often considered "simple" situations, but decisions made during this phase are difficult to change later in the design process. Besides this, the results of global simulation experiments provide the data needed to verify complex simulation models.

The PROPER model has been defined and the use of it during function and process design has been illustrated. However, ttime-dependent behaviour is not included in this model; an extension to it is required to connect the interdisciplinary modelling performed so far to the "world of simulation".

References

Ackoff RL (1971) Towards a system of systems concepts. Manag Sci 17(11):661–671

Bikker H (1995) Organisatie, hoeksteen voor innovatie (Organisation: mainstay of innovation). Inaugural speech, Delft University of Technology, Delft

Burback R (1998) Software engineering methodology: the WaterSluice. Dissertation, Stanford University, Palo Alto, CA

Davenport T (1993) Process innovation: reengineering work through information technology. Harvard Business School Press, Boston, MA

Doepker PE (2001) Integrating the product realization process into the design curriculum. Design Education for the 21st Century 17(4/5):370–374

Dwyer MW (1998) Entelics Research website. http://www.entelics.com/cognitive.html

Jonas W (1997) Viable structures and generative tools: an approach towards designing design. The European Academy of Design, Stockholm

Konda S, Monarch I, Sargent P, Subrahmanian E (1992) Shared memory in design: A unifying theme for research and practice. Res Eng Des 4(1):23–42

Lambert DM (1997) Fundamentals of logistics management. Irwin/McGraw-Hill, New York

Pahl G, Beitz W (1996) Engineering design: A systematic approach, 2nd edn. Springer, London

Pugh S (1990) Total design. Addison-Wesley, Reading, MA

Veeke HPM, Ottjes JA (2000) Objective oriented modeling. In: Proc ICSC Symp Intell Syst Appl 2000 (ISA 2000), Wollongong, Australia, 11–15 Dec 2000

Chapter 11
Case: The Automated Container Terminal

Abstract. The case described in this chapter concerns the design of the first automated container terminal in Rotterdam. The project took place in the early 1990s. This chapter shows the added value of using the Delft Systems Approach in such a design trajectory. It shows that the expectations for the performance of the real system could have been corrected at an early stage of the project if a combination of structure and behaviour models had been used.

11.1 Introduction

In November 1988 a project program was launched to create an innovative container terminal at a peninsula of the Maasvlakte in Rotterdam, the Netherlands, by the main container stevedore in the port of Rotterdam. The growing volume of container transport had prompted the stevedore and a major carrier to reach an agreement about container-handling facilities that would be profitable for both parties. The stevedore was operating in an environment with severe competition, where handling costs and service rates were subject to great pressure. On the other hand, ships and volumes were growing and the carrier had to realise shorter, guaranteed turnaround times on its main shipping lines.

In the years preceding the formulation of the project program, the stevedore performed several feasibility studies, and it was found that a container handling facility could become cost-effective and competitive by automating much of the container handling. In terms of function performance as defined in Chap. 8, the theoretical productivity of this facility was considered to be better than the other alternatives that were investigated.

In this chapter, we describe the design process of the terminal from at the moment the project program appeared. At that time, the stevedore and the carrier had reached an agreement about the functionality and performance of the new terminal. Therefore, the overall strategic phase was finished and the project had entered the tactical design phase. The main goal of this phase was to define a standard productivity for the terminal and all of its components that could be used to evaluate the real operation. To achieve this, all of the project groups involved should have a clear objective in terms of a theoretical productivity with respect to their field of design. In order to assign these productivity figures, the project program of November 1988 is interpreted using the PROcess–PERformance model (PROPER). A global simulation model (TOMAS; Veeke and Ottjes 2000) will be used to interpret these results and to quantify the efforts by means of time-dependent behaviour descriptions.

After that, the use of the approach will be discussed compared with the real execution of the design project.

Note that it is in no way the intention of the analysis presented in this chapter to criticise the former project group and participants (the first author of this book was one of them), even though the outcome of the analysis shows that some of the problems that appeared after the terminal started to operate could have been predicted from this project program. Indeed, the project was a masterpiece of innovative design since it created a completely new container system that "worked" on the intended start date in 1993, and included fundamental innovations throughout the whole system. Proof of this can also be found in practice, because it took almost ten years before real improvements on it could be realised (in Rotterdam and Hamburg).

It should also be noted that computer technology in the early 1990s was not that advanced; complex models required the newest hardware and distributed simulation was not yet being used for this kind of project. Wherever possible, the use of knowledge gained since the original project is avoided here, but even if when is used then the PROPER model is shown to be a suitable method of saving this knowledge in a structured way for future use.

We first provide a summary of the project program (Sect. 11.2), and then the main requirements (Sect. 11.3) are described without comments. After that, these requirements are positioned in the PROPER model and sorted to three system levels: the terminal as a whole, the level of the different aspects (order, product and resource) and finally the function level within the product flow. The requirements for the automated system will then be quantified via a behaviour description and a global TOMAS simulation model.

11.2 The Project Program

The project program was a result of negotiations between the stevedore and the carrier. The decision was made to create a new terminal at the Maasvlakte in Rotterdam that would be fully dedicated to the handling of containers from this one carrier (Fig. 11.1).

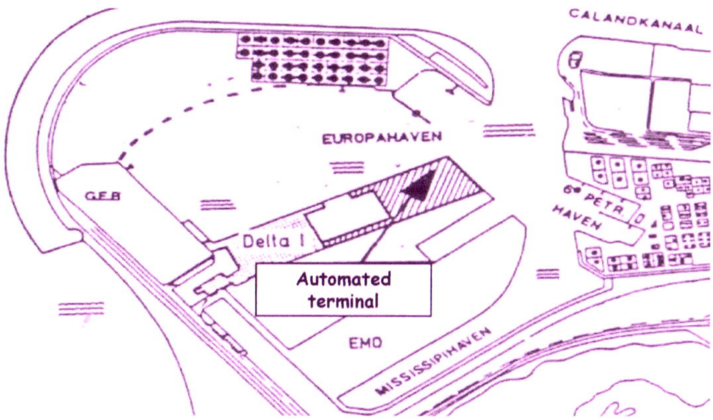

Fig. 11.1 Position of the Automated Container Terminal

The terminal to be built was based on the concept of "majority sea-to-sea" (MSS); it was to be a system for container-handling that was primarily aimed at water–water transfer of containers and was therefore called the "MSS system".

The project organisation (see Fig. 11.2) reflects the multidisciplinary approach taken, as coordinated by a project steering committee in which all project groups were represented.

The "Simulations" group supported all other groups where necessary, and led the development of process control algorithms.

Ideas about the system resulted in the following prescribed combinations of operations and resources (see Fig. 11.3):

- At the waterside, quay cranes (QC) will load and unload deep-sea ships and feeders.
- Automatic guided vehicles (AGV) will transport the containers between stacking area and quay cranes.
- Automatic stacking cranes (ASC) will store and move containers in and out of the stacking area.
- Straddle carriers (SC) will load and unload trucks and multi-trailer systems (MTS) at the land side of the terminal. Each MTS will transport up to five 40-foot containers between a rail or barge terminal and the MSS terminal.

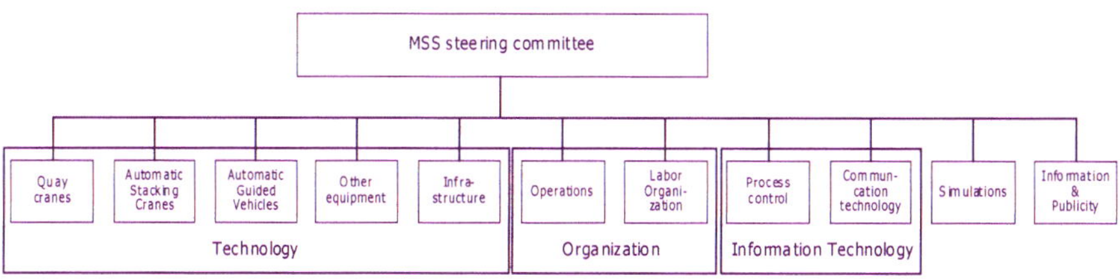

Fig. 11.2 Project organization of the MSS project

Fig. 11.3 An artist's impression of the MSS system (by Rudolf Das)

The MSS system should also be able to cope with expected changes in the market:

- The growing volume of container transfer
- The utilisation of larger transportation units in ocean shipping and in inland shipping and rail, resulting in increased peak-time operations
- A reduction in berth times and service times
- The demand for guaranteed productivity
- The demand for competitive prices compared to other harbours in Western Europe.

These changes were translated into objectives. The main (specifically operational) objectives for the MSS project were:

- To minimise berth times of ships through the optimal use of resources
- To achieve high quay crane productivity
- To increase customer service levels
- To reduce and control costs by increasing productivity, decreasing labour and maintenance costs and increasing the use of information technology.

These objectives resulted in four types of requirements:

- Preconditions that cannot be influenced by the stevedore
- Functional requirements related to the performance, functionality and properties of the terminal
- Operational requirements related to the use of the terminal
- Design restrictions dictated by the organisation.

Preconditions mainly express which system should be used, what types of guarantees should be given and where the terminal should be built; they contain almost no quantitative data. For the purpose of this example the functional requirements are the most important and so these will be elaborated further. Operational requirements and design restrictions will only be mentioned when necessary.

11.3 Functional Requirements

The functional requirements are formulated under the following basic assumptions:

- The source–destination rates of the containers (the so-called *modal split*) are: 55% of the containers imported at waterside will leave at waterside; 38% will leave by rail and road and 7% will leave by barge. 57.5% of the containers leaving at waterside will enter at waterside; 35% will originate from rail and road and 7.5% from barge.
- The arrival pattern of ships will consist of three main line ships together with some tens of feeders each week.
- Ship sizes will be approximately 4000 TEU (twenty foot equivalent unit) for main line ships and approximately 1250 TEU for feeders.
- Average dwell time will be three days for filled containers. The dwell time of a container is defined as the period between unloading it from a ship or truck and subsequently loading it onto a ship or truck.
- The composition of the container flow (full containers, empty containers, reefers and off-standards) will be the same as seen for previous terminals.
- Containers for export will arrive on time and the loading sequence will be known in time in order to guarantee the continuous operation of quay cranes.
- There will be 22 net operating hours each day (24 hours).

The contract with the carrier provides for a review of requirements if these assumptions change. The assumptions are translated into requirements for each aspect, and these should be evaluated regularly.

Finally, it should be mentioned that the contract with the carrier is based on a cost price that encompasses labour costs, materials, third-party services, quay rent and debit and capital costs.

At the level of the terminal as a whole, the following (major) requirements were defined:

- The new terminal should be able to handle 500,000 quay moves per year (including barges). Barges are handled at a separate location. At the sea side of the terminal itself there will be 475,000 moves per year.
- The system as a whole should be able to handle 600 containers a day (24 hours) per quay crane and 630 containers a day after three years.
- The system (excluding the quay cranes) should have an operational reliability of 99%. The average duration of a system failure should not exceed 0.5 hours.

Table 11.1 Length of stay at delivery point

Truck arrivals per hour	Average length of stay	95% level
90–140	30 min	60 min
40–90	20 min	40 min
<40	15 min	30 min

Two general requirements were formulated:

- The number of container movements during stacking and transfer processes should be minimised. This will be achieved by minimising storage layers and using smart stacking control.
- The different container flows will be controlled by a programmed and centralised control.

With respect to the customer's transport modalities, the following requirements were formulated:

A1. As mentioned before, three main line ships per week are expected along with several tens of feeders. In addition to this, the requirements specify the contractual port times for these ships as follows:

- Deep sea: 24–32 hours (depending on the type of ship)
- Feeders: 12 hours

In order to realize these port times an average of 100 containers per gross operational hour (GOH) should be achieved.

A2. The total service time (gate-in/gate-out) for trucks should be approximately 30 minutes. To achieve this, the average length of stay at a delivery position should be 20 minutes. The latter parameter is further specified in relation to the arrival rate (as shown in Table 11.1.)

Single operations (one container) are assumed for the arrivals. For trucks that require more operations (unload and load for example, or two 20-foot containers), the length of stay for an arrival may increase by 50% (independent of the number of operations).

A3. The relations between the arrival and departure of a container were also defined. Import containers from the sea should be available for delivery to the road two hours after unloading. Delivery to rail should be possible three hours after unloading. Export containers can be delivered to the terminal until the ship is moored. In cooperation with carriers, an attempt will be made to have trains and barges unloaded 3 hours before ship arrival.

The project program specifies several requirements for the container flows:

M1. Even under non-optimal conditions, the transfer and stacking system should be able to handle an average of 260 containers each net operating hour (NOH) with eight operational quay cranes.

M2. The stevedore adds the requirement that an average of 40 containers should be handled by each of six quay cranes per NOH. This should result in 240 containers each NOH.

M3. The latter requirement is further refined by the demand that over two consecutive hours one or two quay cranes should be able to handle a peak load of 60 containers per hour. The maximum number of containers to be handled by all six quay cranes each NOH remains 240 in this situation.

M4. At the land side of the terminal, the system should be able to receive and deliver 140 containers per hour (at least during two consecutive hours). Of these containers, 110 containers belong to road transport and 30 to rail transport (which is equivalent to six MTS arrivals). The system should be able to handle 90 containers per hour at the land side.

Besides the requirements for operational circumstances, requirements for maintainability and the intention to use proven technology, the following operational requirements were formulated for the resources:

E1. Eight quay cranes will be provided, three of which should be exchangeable with the neighbouring terminal.

E2. The MSS system will use AGVs, ASCs and straddle carriers.

E3. The MSS system (excluding quay cranes) should have an operational reliability of 99% (with 8760 hours in a year, this means that the system can only be out of action for 90 hours). The duration of failures should be less than 0.5 hours on average and cannot exceed 1 hour.

E4. Each quay crane should have a reliability of 99%. The duration of failures should be less than 0.5 hours on average and cannot exceed 1 hour.

E5. Strategic parts of the system (hard- and software, data communications) should have a reliability of 99.9%.

11.4 Application of the PROPER Model

11.4.1 The Terminal Level

From an organisational point of view, the operation of each terminal is divided into periods of four hours. These "shifts" have a fixed number of operational manned equipment (quay cranes and straddle carriers) and of personnel for process control. Between each shift the current operation status and the responsibility for it is

Table 11.2 The intended results mentioned in the project program

Type	Result	Meas. Unit	Location
Quay moves	475,000	Year	Sea side
Containers (main line)	100	GOH	Sea side
Containers	240/260	NOH	Sea side
Containers	90/140	Hour	Land side
Port times (main line)	24–32 hours	Ship	Sea side
Port times (feeder)	12 hours	Ship	Sea side
Gate-in/gate-out times (truck)	30 minutes	Truck	Land side
Handling times	20 minutes	Truck	Land side
Throughput times	2 or 3 hours	Container	Terminal
Operational hours	22 hours	Day	Terminal

transferred to the new shift. This organisational condition is not mentioned in the project program, but it plays an important role in process control.

In 't Veld (2002) defines three conditions for achieving effective process control:

- There should be a clear objective
- The objective should be feasible
- It should be possible to intervene with the system's behaviour.

No objectives for shifts in terms of intended results and expected efforts are found in the project program; only the intended results per hour, day and year are formulated. By applying the conditions of in 't Veld to the project program along with the restriction of shift-oriented operation, the project program should be interpreted in such a way that at least the intended results and expected efforts for a shift are clear and feasible. Defining interventions is the next step in process design and will not be discussed further in this application. In order be able to intervene effectively, however, the progress of an operation should be clear with respect to the objective, and this factor will be considered here.

Different intended results are defined in the project program. The most significant intended results are summarised in Table 11.2.

In the intended results, the terms "moves" and "containers" are used. At the level of the terminal as a whole, these terms have different meanings: at least two moves are required for each container because it will be stored into the stack by definition. Therefore, at least one move is required to unload and store it and another to retrieve and load it. At the level of the container flow, the terms "move" and "container" are interchangeable: each function within the transfer process handles one container with one move.

The PROPER model (Fig. 5.5) is taken as a starting point for positioning these results. Figure 11.4 shows the result of replacing the general terms of Fig. 5.5 with terms that are specific to this system.

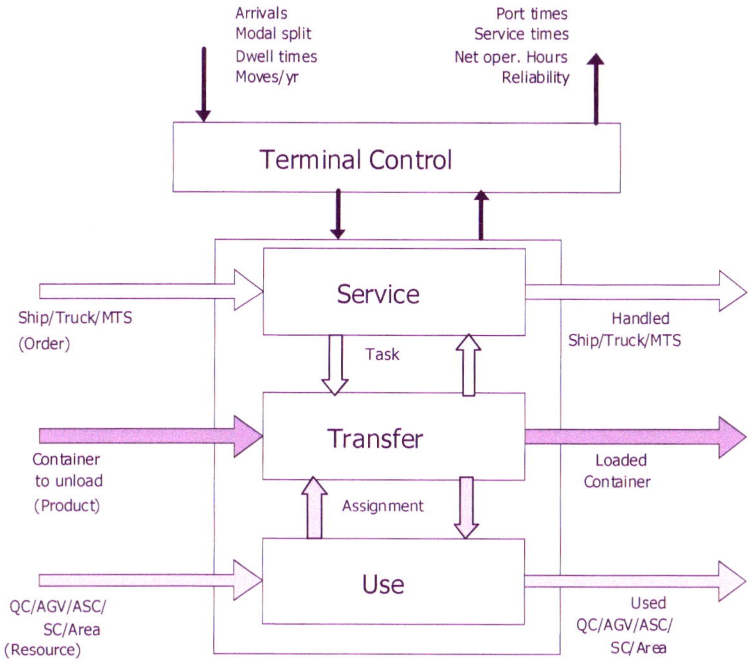

Fig. 11.4 The PROPER model for the terminal

Terminal control reports the performance to the environment, in this case the contractual partner and terminal management. These reports mainly concern the assumptions made and the guaranteed production. These are positioned between the top arrows in Fig. 11.4 because reported results should be formulated in the same terms as the required results.

Upon moving inside the system, these results should be translated into the required results for each flow separately. Results at this level are flow-bound and five flows can be distinguished in the model:

1. Ships, trucks and MTSs are considered *the order flow*. The order process is concerned with the required "customer service". For the modalities ship and truck, the intended results are defined in two ways:

   - Via the port and service times. For MTSs these are currently undefined.
   - Via arrivals (and thus departures) per week.

One condition is added: in order to realise the required port times, 100 containers per gross operational hour (GOH) will need to be handled. This condition is considered to be the first progress indicator of a ship's operation here.

2. Containers represent *the product flow*. This process is concerned with the execution of moves, and it is not necessarily directly related to customer service. Different types of intended results are defined for this flow:

   – *In terms of moves per hour*. Different numbers are formulated: at the sea side 260 moves per hour are required with eight quay cranes and 240 moves per hour with six quay cranes; at land side 90 moves per hour and 140 moves per hour are required with an undefined number of straddle carriers. Quay cranes and straddle carriers operate on both sides of the transfer function; they handle both import and export containers. However, no distinction is made between the input flow rate and the output flow rate (over an hour). It is not yet clear whether this distinction is important.
   – *In terms of quay moves per year*. No land-side figure is defined explicitly, but it can be derived from the assumed modal split. If 55% of the import quay moves represent 57.5% of the export quay moves and the total number of quay moves is 475,000 per year, then there must be around 243,000 import moves and 232,000 export moves per year. Using the percentages of the modal split, the terminal should be able to handle around 209,000 land-side moves per year (truck and rail), divided between 109,000 import moves and 100,000 export moves.

   Beyond this, the project program formulates the ambition to minimise the number of container movements. This defines the intended result that each container will be moved only once by each function in the transfer process.

3. *The resource flow* contains the quay cranes, AGVs, ASCs, straddle carriers and "space" (stack positions, quay length, etc.). The resource process is concerned with technological items, such as reliability, maintenance, safety and operational speed of container moves. In this example, flow rate will not be defined, although the project program mentions their lifespans in terms of operational hours. Operational functions have already been defined for each of them (see Fig. 11.3).

4. *The task flow* between the service function (order flow) and the transfer function (container flow). A task is defined here as one order to store or retrieve one single container. One complete container transit therefore consists of two consecutive tasks, decoupled by the ASC stack. A task's lead time is a measure of the response time of the system to container orders, which results in a required service time or port time; by adding priority levels to tasks the service function is able to intervene with the progress of sea-side and land-side operations.

   The project program defines the moments when tasks should be ready for release; e.g. a truck with a container that needs to be loaded into a ship should be present before the mooring time of the ship. The intended results for this flow are expressed in terms of the number of tasks per shift and the task lead times per shift.

This example clearly illustrates the difference between order flow and (internal) task flow. One ship (a customer order) results in a large number of separate tasks. This introduces a number of choices to be made. One could decide to release all tasks immediately, but this neglects the formulation of a shift operation objective. One could also decide to release tasks one-by-one, but this could introduce delays into the operation and exclude the ability to optimise task scheduling.

5. *The assignment flow* between the use function and the transfer function. The intended results of assignment are not defined in the project program. They should be expressed in equipment moves per shift, availability and occupation per shift, because manned equipment will be assigned per shift. These results form the basis for calculating the expected efforts at the terminal level.

The task and assignment flows contribute to insight into the factors that influence an operation's performance. A bad performance can be the result of a late task release, long task lead times or a shortage in assigned equipment.

The use of shift-based operation forces the intended results of the project program, as formulated in different time units, to be translated into intended results per shift. Some intended results are even missing. In order to define all of these results, a step that involves zooming into the model of Fig. 11.4 will be performed, resulting in separate models for each of the horizontal flows.

11.4.2 The Order Flow

Two geographically divided flows are distinguished at the terminal (Fig. 11.5):

- A sea-side flow consisting of ships and feeders
- A land-side flow consisting of trucks and multi-trailer systems (MTS).

Multi-trailer systems (MTS) can transport up to five 40-foot containers. They transport containers to and from trains, barges and other terminals in the area. The MTS transports are not the direct responsibility of the ACT terminal and are therefore considered a "customer" modality.

Tasks should be divided into tasks for the sea side and land side, and into tasks to store and retrieve a container. By measuring the performed tasks per GOH, the control service function is able to judge the progress of the service. The value is transformed into performed tasks per shift by simply adding the values of the four shift hours (a shift is in fact a period of four GOHs). The measured task lead times per shift and the number of tasks released per shift enable the control service function to derive the standard results per shift for different arrival compositions and ship operation phases. The period between the unload task and the load task of a specific container represents the dwell time.

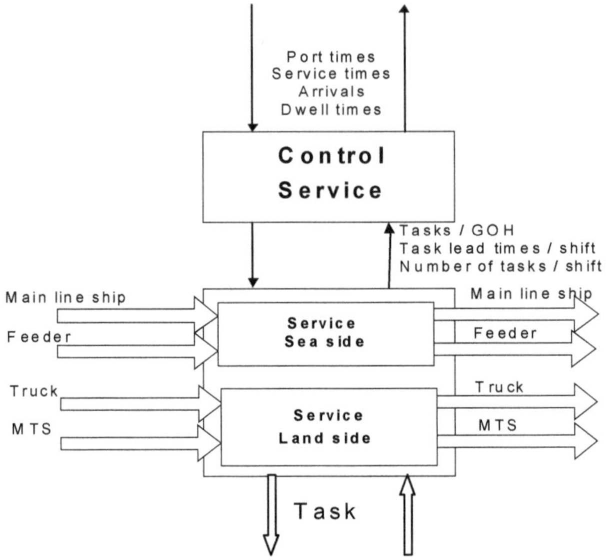

Fig. 11.5 The order flow at the ACT terminal

11.4.3 The Product Flow

In the container flow, two geographically separated flows can again be distinguished. The flows exhibit overlap in the ASC stack (Fig. 11.6).

The equipment used for each function is shown in the bottom-right corner of the function. On the land side, the straddle carriers perform the functions "unload", "transport" and "load".

The control transfer function combines the released tasks with the assigned resources to perform the transfer function. The goal is to achieve the required tasks per shift with the resources assigned to the operation. The tasks will be split into separate tasks for each function. Consecutive functions within the flow should be executed in an optimal way.

Handled tasks are returned to the service function and assigned resources are released at the end of a shift.

The intended results as formulated in the project program are expressed in terms of "net operational hours". No explanation of how this unit is measured is provided. By just measuring the moves per hour (which is in fact a gross operational hour), the relation to the number of moves per year is made clear and it can be used by terminal control as a progress indicator during the year.

The ASCs are the only resources involved in both the sea-side operation and the land-side operation. This means that the effective operational capacities of the sea-side system and the land-side system fluctuate during a shift. Whereas the

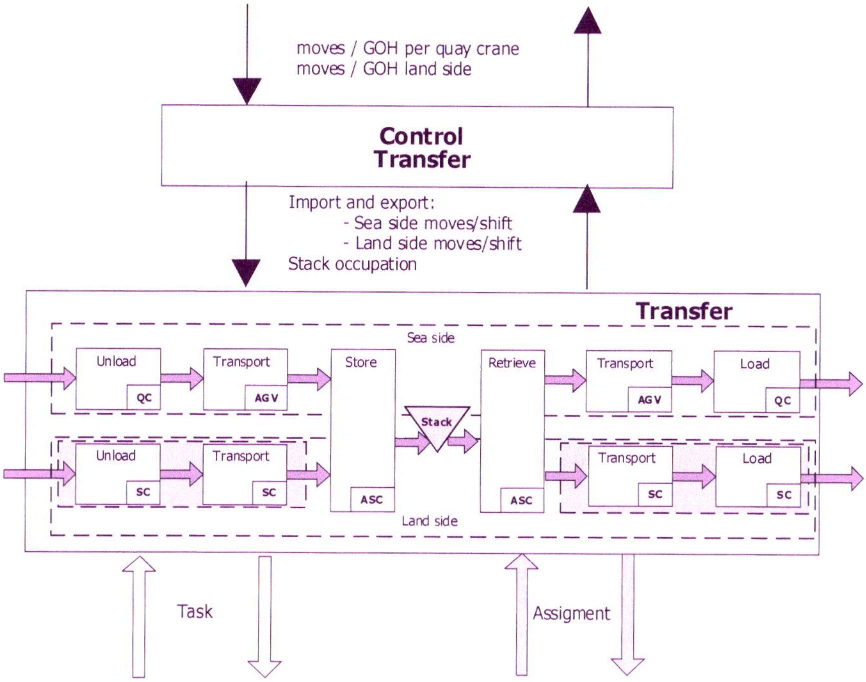

Fig. 11.6 The container flow at the ACT terminal

other resources can be assigned to the sea side or the land side exclusively, the ASC resources need to be assigned to the sum of sea-side and land-side tasks. Since land-side tasks are often unpredictable (and so are the containers to be requested are too), this is an extra reason to formulate intended results in terms of shift units rather than hour units.

Besides this, each ASC is restricted to its own part of the stacking area. This is no problem for the unload flows to the left of the ASC stack in Fig. 11.6: the control function is free to choose an ASC area at will. However, it is an extra complication for the load flows to the right of the ASC stack. In trying to optimise move times, the control function is bound to the ASC area, where the container is stored. This may lead to task congestion for a specific ASC, influencing task lead times. It is therefore concluded that a distinction should be made between input moves and output moves.

Each of the functions in Fig. 11.6 should have the same intended result defined in order to prevent the creation of a bottleneck. In particular, the sea-side operation consists of three closely coupled functions with different resources. If the intended results have been defined, then the expected costs for each of the resources should be derived. This will be investigated further in Sect. 10.5 using simulation.

11.4.4 The Resource Flow

The use function aims to provide the required resources for the operation function. Manned resources (quay crane and straddle carrier) and personnel are assigned on a shift basis. There is no reason to use another time basis for the automatic resources. The results of assignment will therefore be expressed in terms of results per shift. The project program only defines results for assigned resources in terms of reliability. It is assumed here that 100% availability for assignment is required. Two things are required to achieve this:

- The amount of operational resources available should be sufficient. The maintenance program and failure rates should allow this.
- The amount of required equipment should enable the considered operational results to be achieved. The technical specifications should allow the operational requirements as mentioned in the preceding section.

Assigned resources make up the labour and capital costs of the operation and form the basis for calculating the expected efforts. They will therefore be measured and reported to the terminal control function.

The use function obtained after restricting it to equipment and skipping personnel is represented in Fig. 11.7.

During its lifetime, the equipment will "loop" between the status "usable" and the status "in maintenance". Inside the use function, these loops should be controlled in order to guarantee the required availability, which is an intended result that is used externally.

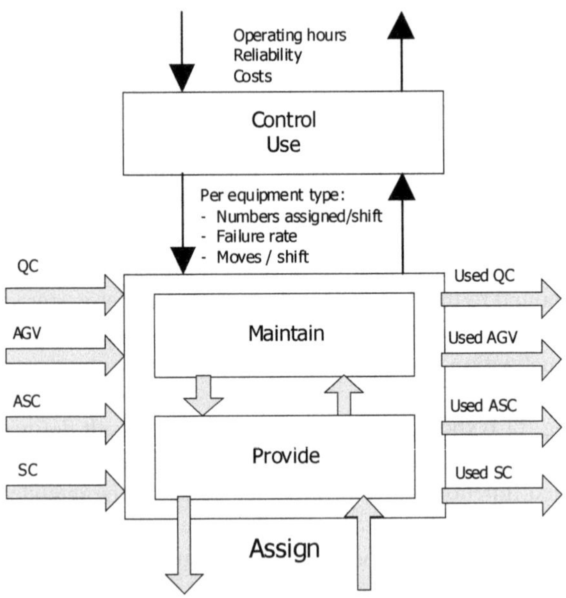

Fig. 11.7 The use function of the ACT terminal

11.5 Behaviour Descriptions for Productivity Definitions

11.5.1 Goals of the Behaviour Descriptions

The availability requirements (and thus the expected efforts) were not made explicit in the original project program. Only the intended results for quay cranes were defined, but with different time units than required for the control functions. Task lead times, transport, store and retrieve times still have to be defined, as shown in the previous paragraphs. These data are required to answer questions about availability and to define technical requirements for the equipment. Using these data it will be possible to judge the feasibility of the defined intended results. When the project program was created these data were unknown, because the ASCs and AGVs still had to be developed from scratch. Simulation can however be used to derive these data at the level of the functions of Fig. 11.6. Assuming that the quay cranes are able to achieve the required performance, then the requirements should be translated into requirements for the AGVs and ASCs.

For illustration purposes, the next study is restricted to the import functions and within these the operational processes of containers, and the land-side operation is taken into account up to the point at which a container is retrieved from the stack. The resources used for each function are added, thus revealing their repetitive processes. The resulting process is shown in Fig. 11.8.

The triangles in Fig. 11.8 represent waiting times. The shaded triangles represent "idle times" for the resources: they are waiting for a task to be assigned. The white triangles are waiting times caused by imperfect synchronisation.

It is assumed that the assignment of resources to the operation has been taken care of and that all quay cranes, AGVs and ASCs are available. In the following description, it is assumed that tasks are known and are available in time.

During each "container move", two functions are performed for each piece of equipment: a "move" to the position to get a container and then the performance of the actual function: "unload" (QC), "transport" (AGV), "store" and "retrieve" (ASC), which is finished by releasing the container. We call the performance of

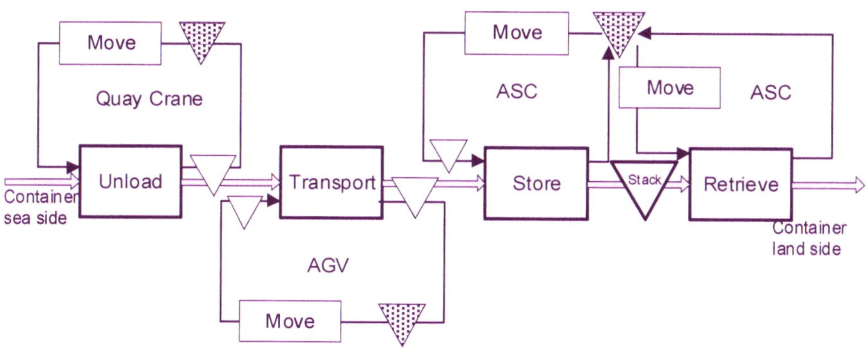

Fig. 11.8 The import process of the ACT terminal

these two functions together a "cycle" from now on. AGVs do not have a facility to lift or put down a container, and so their operation is strongly linked to the quay cranes and ASCs. The consequence is that a third factor should be accounted for in the definition of a cycle: the transfer of a container between the AGV and the other equipment. The technical cycle time of equipment can be defined with these factors.

At the land side the situation is different, because there are buffer positions where the ASCs and straddle carriers can pick up or put down containers independently.

In order to formulate the equipment requirements for a feasible result for the system, the following global process description for an unloading process is constructed (Table 11.3). Lines that start "Q*n*" refer to questions that are defined after the table.

In this case every type of equipment (except a straddle carrier) has a behaviour description, a process. The model assumes undisturbed operation: there is always a sea-side job for a quay crane. The intended results that are used are therefore the results per net operational hour of the project program. At the land side it is assumed that 90 containers per hour are handled (jobs arrive at a constant rate).

Table 11.3 Process descriptions for the MSS system

Process for a Quay Crane	Process for an AGV
Repeat Q1 Work "sample of QC's cycle time" Wait while no AGVs waiting Q2 Select AGV Q3 Load AGV in *x* seconds Resume AGV's process	Repeat Q4 Select QC Q5 Cycle is "sample of AGV's cycle time" Drive 0.5 × Cycle* Q6 Wait for QC Q7 Select ASC Drive 0.5 × Cycle* Q8 Wait for ASC
Process of an ASC	**Process of Landside**

Repeat
Wait while no AGVs waiting
and no jobs on land side
Q9 Cycle = sample of ASC cycle time
Q10 If AGVs waiting
Q11 Select AGV
Move to AGV in 0.5 × Cycle*
Q12 Unload AGV in *y* seconds
Resume AGV's process
Store container
in 0.5 × Cycle*
Else
Select land-side job
Work *D* seconds
	Repeat
	Wait(3600/90)
	Q13 Select ASC
	Create land-side job for ASC

* The time taken to move to the position where the AGV or ASC can get the container is assumed to be half the cycle. This does not affect the results.

Straddle carriers are not modelled because the transfer between the straddle carrier and the ASC is decoupled using a buffer. Only the workload resulting from the straddle carriers is modelled. The cycle times mentioned in Table 11.3 exclude the transfer time at the transfer position.

The goal of the model is to gain insight into feasible productivity, not to optimise the operation. Control algorithms are not modelled. Bearing this in mind, the process descriptions introduce 13 questions to the project team:

Q1. What is the cycle time of a quay crane's unload move?
Q2. How does a quay crane select an AGV from the AGVs waiting to be loaded?
Q3. How long does it take for a quay crane to load an AGV (x seconds)?
Q4. When and how is the AGV that the quay crane drives to selected?
Q5. What is the cycle time of an AGV, where a cycle is the total time taken to drive to a quay crane and return to an ASC?
Q6. Is there an upper limit on the number of AGVs waiting for a quay crane?
Q7. How does an AGV select an ASC to deliver the container to?
Q8. Is there an upper limit on the number of AGVs waiting for an ASC?
Q9. What is the cycle time of an ASC?
Q10. Is there an order of priority between sea-side and land-side jobs that need to be handled by the same ASC?
Q11. How does an ASC select an AGV if there is more than one AGV waiting?
Q12. How long does it take for an ASC to unload an AGV (y seconds)?
Q13. How are ASCs selected for land-side jobs?

The amount of equipment together with their cycle times (Q1, Q3, Q5, Q9, Q12) mainly determine the results at the sea side of the system. The most likely or the "best-case scenario" can be chosen for the others (for the time being).

The (assumed) answers to the other questions are now:

Q6, Q8: There are no upper limits on the number of AGVs waiting for a QC or ASC.

Q2, Q11: Both quay cranes and ASCs select AGVs in first-in-first-out order. This selection does not influence the results at this level of detail; neither AGVs nor containers have an identity here.

Q10: The process description shows that sea-side jobs are selected before land-side jobs, because an ASC first checks if AGVs are available. This will maximize the sea-side performance.

Q7, Q13: Both AGVs and land side select an ASC according a uniform distribution.

A second alternative could be investigated where AGVs and land side select an ASC according the work load already assigned to an ASC. This selection is assumed to be performed at the moment of arrival.

One should bear in mind that workload-based selection is only possible for import containers. Export containers are already in the stack and thus the ASC to handle them are determined beforehand.

Q4: This question is highly influential, because quay cranes are highly unpredictable. If an AGV selects a quay crane early on in its cycle then this can be a bad choice if the quay crane randomly exhibits large cycles after this choice.

For the "best-case scenario", AGVs do not select a quay crane at all. They are assumed to arrive in a large pool of waiting AGVs, where each quay crane can select an AGV instantaneously. In this alternative, the consequences of early (and potentially non-optimal) selection are excluded. A second alternative will be the situation where an AGV selects the destination quay crane at the moment of arrival in the quay crane area. Finally, in the third alternative an AGV selects a quay crane at the moment of departure from the ASC.

Reality will lie somewhere between the second and third alternative. The first alternative will only be used to find a starting point for other experiments.

The following assumptions will be made:

- A quay crane will load an AGV in 15 seconds.
- QC cycle times for unloading will be 95 seconds in cases with eight quay cranes and 75 seconds in cases with six quay cranes. Including the loading time of AGVs, this means a (maximum) average production of 260 containers per hour for eight quay cranes, and a production of 240 containers per hour for six quay cranes. The 260 and 240 containers per hour agree with the requirements mentioned in the project program.
- An ASC will unload an AGV in 15 seconds.
- The cycle times of the AGVs and ASCs are assumed to be uniformly distributed. The cycle times will be the result of the (still unknown) routing algorithm to be used as well as traffic delays.
- The cycle times of the quay cranes are assumed to have a negative exponential distribution.

A simulation model has been written based on the description given in Table 11.3, in which the number of quay cranes, AGVs and ASCs and the cycle times can be varied.

10.5.2 Experiments

In the model, 48 AGVs and 25 ASCs will be assumed to be available. These numbers were used during the early stages of process design. All of the experiments below can be repeated with other numbers for cases with different availability figures. As long as these numbers are accepted, they become a standard result for the use function of Fig. 11.8.

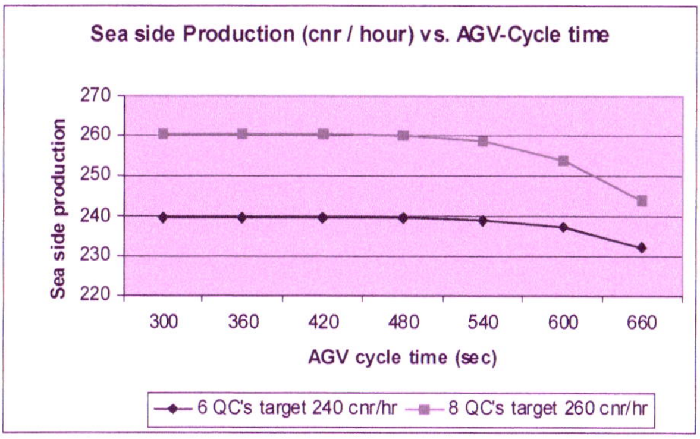

Fig. 11.9 Sea-side production as a function of AGV cycle time

Each experiment covers a period of 100 consecutive net operational hours, after an initial hour to start operations.

The first experiments are meant to provide an indication of the cycle times required to achieve a sea-side productivity of 260 cnr/NOH with eight quay cranes and 240 cnr/NOH with six quay cranes. First the required cycle time of the AGVs is determined by setting the cycle time of the ASCs to zero. The cycle times of the AGVs tested (including transfer times) are 300, 360, 420, 480, 540, 600 and 660 seconds. The results are shown in Fig. 11.9.

Figure 11.9 shows that a maximum technical cycle time of 480 seconds is allowed for both eight-quay-crane and six-quay-crane operations. A longer cycle time will cause the AGV system to become a bottleneck in the system, decreasing the resulting productivity.

In the next experiments the ASC cycle time (including the transfer time) is gradually increased. The cycle times tested are 60, 120, 180 and 240 seconds. This will also influence the result for the AGV system, because AGVs may now be waiting for an ASC. For this reason, the experiments are performed with AGV cycle times of 360, 420 and 480 seconds. The results are shown in Fig. 11.10.

From Fig. 11.10, it can be concluded that the required production is still feasible if the average ASC cycle time is around 120 seconds and the AGV cycle time is less than 420 seconds. An ASC cycle time of 180 seconds combined with this AGV cycle time may be acceptable if process control is able to make use of random advantageous circumstances (for example by a partial overlap of the ASC move function and the AGV transport function; in this case the operational cycle time of the ASCs actually decreases).

Upon reviewing the control alternatives derived from Table 10.2, six experiments were defined, as shown in Table 11.4.

Experiment A1 has already been executed; it will serve as a reference experiment, and the other experiments will be executed with the following average cycle times: 180 seconds for an ASC and 360 seconds for an AGV. The results are shown in Fig. 11.11.

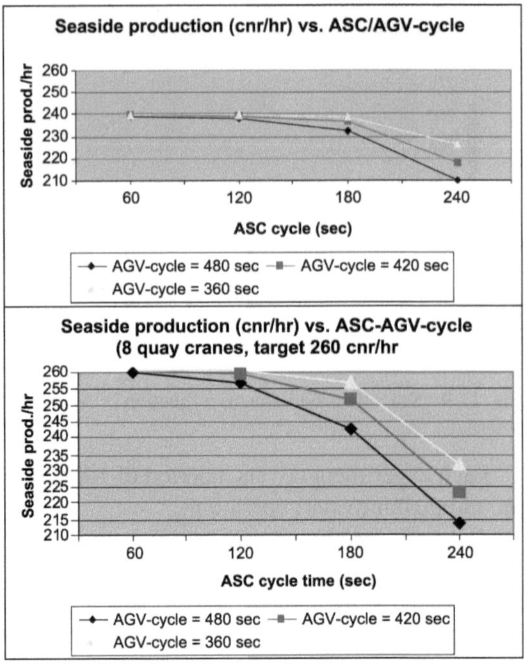

Fig. 11.10 Sea-side production as a function of AGV and ASC cycle times

With an ASC selection based on the workload, the graphs show that sea-side productivity is less sensitive to the moment of QC selection. This workload-based ASC selection is however only possible for an unloading operation. During a loading operation the loading sequence determines both the ASC and the QC to be selected. Task scheduling can only influence the selection effectiveness by selecting an appropriate quay crane. However, quay crane selection must then be performed even earlier than in the cases A3 and B3: at the start or during the move function of an AGV. To get an impression of the effects of this, a final experiment was performed where experiment A3 was adapted for loading operations in such a way that the AGVs select a quay crane at the moment of departure from a quay crane. The experiments show that the production decreases to 235 and

Table 11.4 Experimental scheme

Experiment A: Uniform distributed selection of ASCs
A1. No selection of QC by AGV A2. Selection of QC by AGV upon arrival at the quay A3. Selection of QC by AGV upon departure from the ASC
Experiment B: Selection of ASCs based on work load
B1. No selection of QC by AGV B2. Selection of QC by AGV upon arrival B3. Selection of QC by AGV upon departure from the ASC

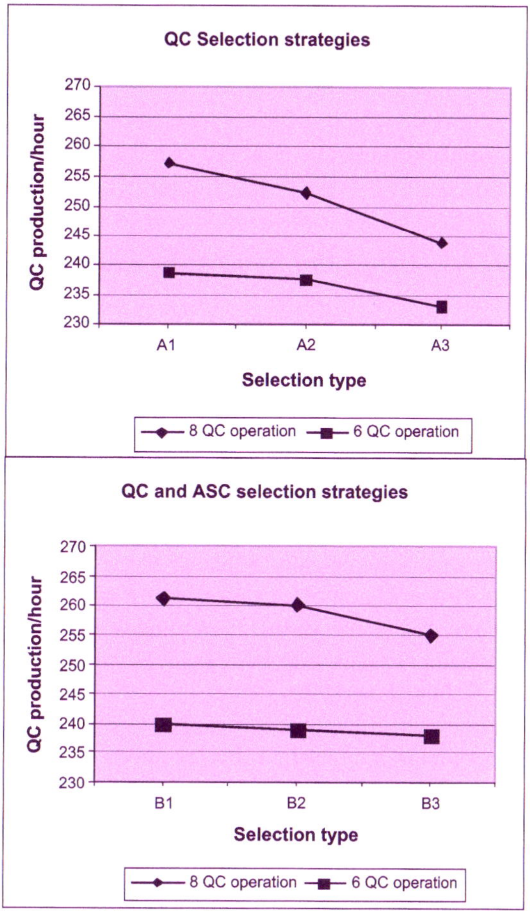

Fig. 11.11 The effects of QC and ASC selection strategies

227 cnr/NOH for operations with eight and six quay cranes, respectively. Further decreases should also be expected as a result of sequencing problems.

11.5.3 Results

The conclusions from the preceding experiments are (given the intended results for net operational hours in the project program):

- The AGV system should be capable of delivering an average "technical" AGV cycle time of around 360 seconds, including transfer times at the QCs and the ASCs.
- The ASC system should be capable of delivering an average ASC cycle time of around 180 seconds, including transfer times at the transfer positions.

- The required productions are assumed to be productions of unloading operations.
- The cycle times of the quay cranes are assumed to be technical cycle times, including transfer times.
- An ASC selection strategy based on workload has a positive effect on production and should be used to make the requirements feasible.

11.5.4 Using the Behaviour Model During the Project

The process descriptions of Table 11.3 form an appropriate basis for discussing these requirements at an interdisciplinary level. The requirements can be translated directly into a computer simulation model. Using this approach, the intended results can be quantified up to function level, and they should be used by the project groups as objectives for detailed design. For example, the AGV cycle times are objectives for the project groups that develop the AGVs and traffic control. If a (technological) driving speed of AGVs is decided on during some design phase, then this automatically leads to a maximum allowed distance per AGV cycle, which then becomes an objective for process control. The same conclusion holds for the ASC cycle times. If at some moment during process design these requirements do not appear to be feasible, then the consequences for other groups should be evaluated using this model.

In reality, at some point the technical specification of the AGVs defined a maximum straight-line driving speed of 3 m/s and 2 m/s around curves. At that moment no traffic routing mechanism had been decided upon and the layout of the quay area suggested that there was no reason to suspect that the cycle time of 360 s would not be feasible. In a later phase, however, the routing algorithm was defined and this showed long driving distances, usually with six 90° curves in each cycle. In reality, the measured maximum sea-side production of terminal operations was around 210 containers/GOH for unloading operations.

If the technical data on routing distances and AGV speeds had been applied to the global behaviour model, this production could have been predicted. A raw calculation shows that the technical cycle time would be around 420 seconds and traffic delays had not yet been included. With this knowledge, experiment B3 could have been re-executed with an average AGV cycle time of 480 seconds. The results would have shown a sea-side production of 228 cnr/NOH with six quay cranes. If the condition of 22 NOH per calendar day (24 hours) is applied, the measured value of 210 cnr/GOH matches the simulation results.

During the project several detailed models were developed, for example an AGV traffic control model, an ASC system model and a straddle carrier model. These were all standalone models, with assumed inputs at their system borders. It is now possible to connect each of these models to the global model described above using the send and receive mechanism described in Chap. 6. By applying the object-oriented approach in connection with the process interaction approach, an object class (for example the AGV or ASC system) can be extracted from the global model and implemented to any level of detail. Each of these models makes use of the input flows as they are generated by the global model. In this way, the results from detailed designs can be compared with the high-level requirements.

Finally, the same approach can be used to test the real system with respect to control algorithms (task scheduling, traffic control) and equipment.

11.6 Conclusions

In this chapter, the interdisciplinary use of the PROPER model and behaviour descriptions in a complex design project has been illustrated. They were applied to the results from the function design phase (as they are formulated in the project program).

It has been shown that the PROPER model structures the theoretical productivity values presented in the project program. It thereby relates the values for service, operation and technology to each other and to the overall terminal values. It enables the intended results to be defined for all project groups, starting with process design. By using the PROPER model, values that are insufficient are detected and the suitability of other values can be reviewed; for example, not all of the intended results for shifts were defined and no availability requirements were formulated.

It was also shown that time-dependent behaviour can be studied at a global level of detail, as an aid to quantifying defined objectives and deriving objectives for service, transfer and resource use. The global descriptions can be detailed further according to the function structure of the PROPER model. By doing this, the original requirements and product and order flow characteristics are preserved, and adjustments can be made as soon as original objectives become infeasible.

In reality, this systematic analysis and behavioural approach was not applied; driving speeds were selected on a technological basis only, and simulations and process control solely tried to maximise production. The AGV traffic control had to be developed from scratch. Initial ideas included "free-ranging" AGVs (with short driving distances), and it took some time before a strict (and apparently non-optimal) routing control for AGVs was decided on. This routing control resulted in driving distances that—considering the maximum driving speed—could not fulfil the requirements.

Therefore, real production remained below the requirements formulated in the project program, mainly because the expectations were not realistic.

Although it cannot be proven that applying the approach described in this chapter would have improved the real results, we have shown here that it would have clarified expectations at an early stage, and so proper measurements could have been taken to improve matters.

References

Veeke HPM, Ottjes JA (2000) Objective oriented modeling. In: Proc ICSC Symp Intell Syst Appl 2000 (ISA 2000), Wollongong, Australia, 11–15 Dec 2000

in 't Veld J (2002) Analyse van organisatieproblemen, een toepassing van denken in systemen en processen (Analysis of organisation problems, an application of thinking in terms of systems and processes). Stenfert Kroese, Leiden

Index

MIX
Papier aus verantwortungsvollen Quellen
Paper from responsible sources
FSC® C105338

If you have any concerns about our products,
you can contact us on
ProductSafety@springernature.com

In case Publisher is established outside the EU,
the EU authorized representative is:
Springer Nature Customer Service Center GmbH
Europaplatz 3, 69115 Heidelberg, Germany

Printed by Libri Plureos GmbH
in Hamburg, Germany